# EIGHT DYNAMIC PATTERNS OF LIVING

## Essays on Law, Policy and Psychiatry
## Peter Fritz Walter

---

### Codependence
Coping with Addiction, Sadism and Abuse

### Eight Dynamic Patterns of Living
Base Elements of True Civilization

### Emotional Flow
A Holistic Approach to Healing Sadism

### Love or Laws?
When Law Punishes Life

### Minotaur Unveiled
A Historical Assessment of Adult-Child Sexual Interaction

### Natural Order
Thesis, Antithesis and Synthesis in Human Evolution

### Pedophilia Revisited
The Making of a Crime for Justifying Lacking Social Policy

### The Commercial Exploitation of Abuse
A Study on Social Policy

### The Legal Split in Child Protection
Overcoming the Double Standard

### The Roots of Violence
Why Humans Are Not by Nature Violent

# EIGHT DYNAMIC PATTERNS OF LIVING

## Base Elements of True Civilization

Peter Fritz Walter

Published by Sirius-C Media Galaxy LLC

Business Filings Incorporated

108 West 13th St., Wilmington, DE 19801, USA

©2018 Peter Fritz Walter. Some rights reserved.

Essays on Law, Policy and Psychiatry, Vol. 3

Creative Commons Attribution 4.0 International License

This publication may be distributed, used for an adaptation or for derivative works, also for commercial purposes, as long as the rights of the author are attributed. The attribution must be given to the best of the user's ability with the information available. Third party licenses or copyright of quoted resources are untouched by this license and remain under their own license.

The moral right of the author has been asserted

Set in Palatino

Designed by Peter Fritz Walter

ISBN 978-1-983988-30-1

Publishing Categories
Science / Life Sciences / Evolution

Publisher Contact Information
publisher@sirius-c-publishing.com
http://sirius-c-publishing.com

Author Contact Information
pfw@peterfritzwalter.com

About Dr. Peter Fritz Walter
http://peterfritzwalter.com

## About the Author

Parallel to an international law career in Germany, Switzerland and the United States, Dr. Peter Fritz Walter (Pierre) focused upon fine art, cookery, astrology, musical performance, social sciences and humanities.

He started writing essays as an adolescent and received a high school award for creative writing and editorial work for the school magazine.

After finalizing his law diplomas, he graduated with an LL.M. in European Integration at Saarland University, Germany, in 1982, and with a Doctor of Law title from University of Geneva, Switzerland, in 1987.

He then took courses in psychology at the University of Geneva and interviewed a number of psychotherapists in Lausanne and Geneva, Switzerland. His interest was intensified through a hypnotherapy with an Ericksonian American hypnotherapist in Lausanne. This led him to the recovery and healing of his inner child.

After a second career as a corporate trainer and personal coach, Pierre retired in 2004 as a full-time writer, philosopher and consultant.

His nonfiction books emphasize a systemic, holistic, cross-cultural and interdisciplinary perspective, while his fiction works and short stories focus upon education, philosophy, perennial wisdom, and the poetic formulation of an integrative worldview.

Pierre is a German-French bilingual native speaker and writes English as his 4th language after German, Latin and French. He also reads source literature for his research works in Spanish, Italian, Portuguese, and Dutch. In addition, Pierre has notions of Thai, Khmer, Chinese, Japanese, and Vietnamese.

All of Pierre's books are hand-crafted and self-published, designed by the author. Pierre publishes via his Delaware company, Sirius-C Media Galaxy LLC, and under the imprints of IPUBLICA and SCM (Sirius-C Media).

The author's profits from this book are being donated to charity.

# Contents

| | |
|---|---|
| **Preface** | 9 |
| A Common Etiology | |
| | |
| **Introduction** | 21 |
| | |
| **The Eight Patterns (Overview)** | 31 |
| 1) Autonomy | 31 |
| 2) Ecstasy | 32 |
| 3) Energy | 33 |
| 4) Language | 35 |
| 5) Love | 37 |
| 6) Pleasure | 38 |
| 7) Self–Regulation | 39 |
| 8) Touch | 39 |
| | |
| **The Autonomy Pattern** | 41 |
| | |
| **The Ecstasy Pattern** | 51 |
| | |
| **The Energy Pattern** | 67 |
| | |
| **The Language Pattern** | 73 |
| | |
| **The Love Pattern** | 87 |
| Culture and Pleasure | 87 |
| Pleasure–Denial and Violence | 88 |
| Compulsory Sex Morality | 89 |

| | |
|---|---|
| Anthropological Evidence | 94 |
| Love Osmosis | 98 |
| Love versus Morality | 100 |
| Rebuilding Trust | 107 |

## The Pleasure Pattern — 111

## The Self–Regulation Pattern — 141

## The Touch Pattern — 163

## Bibliography — 181

## Personal Notes — 243

# Preface

## A Common Etiology

In this essay I shall first start with an introduction into systems theory, and the sketch of a critique of traditional Newtonian-Cartesian 'clockwork' science. We are going to see that I consider the shift from Cartesian science to present–day holistic science as a tremendous upward-lift of human consciousness.

I found *eight dynamic patterns* in the lifestyle of native peoples around the world, which are autonomy, ecstasy, energy, language, love, pleasure, self–regulation, and touch. As to our own culture, the lack of recognition of the eight patterns of living equals *eight huge black holes* in the cultural setup of modern society, which is the underlying and hidden reason of our social chaos, rampant violence and almost total disappearance of integrative social values.

# EIGHT DYNAMIC PATTERNS OF LIVING

The *eight dynamic patterns of living* have to this day never been integrated by Western society, but the one pattern that really clashes with our scientific materialism is *energy*, not energy as kinetic energy, of course, but *energy as cosmic, vibrant subtle energy*, the all–permeating bioplasmatic energy, the information field, the creator force, the cosmic and human energy fields, the *orgone*, as Reich termed it.

Wilhelm Reich (1897-1957) was only one among a number of pioneers who questioned the status quo of Cartesian rigidity and introduced an energy-based science paradigm, while they gave different names to this concept; these men were attacked and persecuted by the *system-conform* science establishment. Today, while most of this controversy is settled and the information field recognized as the *zero-point field, unified field* or *quantum vacuum*, the seven other dynamic patterns are not for that matter recognized and embraced by postmodern international consumer culture.

A detailed regard on each pattern, with the *autonomy pattern* on top of the list, will show that autonomy is fundamental for every being in growth, and that the absence of autonomy within our modern

## PREFACE

child-rearing paradigm brings about *rampant codependence and emotional abuse* in the parent-child relationship, and within the modern nuclear family. Considering anthropological field research and cross-cultural data, we see that if the child is deprived of the autonomy to engage in erotic peer relations, as this is the case with the emotionally abusive Western child-rearing paradigm since about the 17th century, the natural balance within the family gets upset; parents, then, develop an erotic, *pedophilic*, attraction toward their children, and children develop a *gerontophilic* fixation upon their parents and caretakers. These fixations are thus not instituted by nature, as some psychoanalysts wrongly assumes, but are the *result of Western children's deprivation of autonomy*, as a result of overprotection and the societal prohibition for the child to be actively sexual, in partner relations, and not just merely auto-erotic and masturbatory.

*Ecstasy*, the second of the eight dynamic patterns of living, is not to be confounded with pleasure, and does not represent a form of sexual licentiousness. According to Terence McKenna and other shamanism researchers, ecstasy is a 'contemplation of wholeness,' a vision of unity that is truly a form of direct religious experience. This consciousness pattern is almost entirely lost within modern consumer culture, which is well hedonistic in the sense of pleasure-affirming, but not interested in the numinous, religious experience as a direct source of personal growth and rejuvenation.

The *energy pattern* is existent without exception in all tribal cultures, but was strongly, over centuries, banned from Western culture as a *religious heresy* and a scientific monstrosity, which led to the persecution and in some cases the death of scientists (like Giordano Bruno) who were interested to find the secret of the 'forbidden tree of knowledge.' However, when we look at cutting edge physics, and research on the quantum field, we can say that the taboo is presently being lifted and the energy view of the world by and by integrated within Western science, over the course of about the last two decades, and under the pulpit of *quantum physics*.

When we look at the *language pattern*, we see that *culture and language* are one interdependent whole, and that there is no culture without language and no language that is not imbedded in a cultural

# PREFACE

context. This means, as a consequence, that, according to psychoanalyst Robert M. Stein, the imaginal realm belongs to the cultural achievements and is therefore sacred and exempt from being controlled by sociopolitical and moral rules and laws.

This is the case with native cultures where all taboos are *do-taboos*, not talk-taboos, but it is not so, and was not so historically, within Western culture, where the Church demanded absolute control also over the inner world of the believer.

In our culture, rule and exception seem to be reversed here, in that what is not talked about, and is not sacralized through ritual and prayer, because it's a *non-dit*, is done in secret, while it is vehemently attacked and declared as vicious and abject in the public debate. Examples of matters that are blinded out and tabooed within modern culture are *parent-child incest and pedophilia*, which is why these modes of behavior are not socially coded and integrated, but are creating havoc in that without a valid social code, they represent chaotic forms of human conduct.

The reason is, as we saw it, that they are not being embraced as valid forms of human behavior *on the imaginal sphere*, and therefore are not openly discussed in films, books or in television—which would lead to their being embedded in *culture* and thus, would hold them within the limits of the culture. As they are *not embraced but disintegrated*, they became rampant, uncontrolled and harmful on the social scale.

The primary reason, to repeat it, for this to happen is that the language pattern has been shunned by Western society through hundreds of years of Church terror and speech taboos of all sorts. That freedom of speech is now a constitutional guarantee has *not changed this fact* because of the widespread censorship being in place regarding the tabooed and non-coded behavior modes that are rejected by the culture, and banned from publishing on a large scale, both in book and in online publishing.

Hence, despite a modern constitution, the reality in a culture that has disintegrated the imaginal realm, and the language code to express it, is that contrary to the official rhetoric, censorship is the order of the day and the rule, while free speech is the exception.

The *love pattern* is to be found with all native cultures, in the sense that their overall credo could be

## PREFACE

expressed in the slogan 'life is love,' while the overall credo of modern culture would sound like 'life is power' or 'life is control.'

Tribal cultures that have an intact love pattern in place, have basic trust in life. Regarding the love pattern, we can make out another rule-and-exception principle, in the sense that trust was originally the rule with human societies, while in modern culture, the rule is *mistrust*. The fear of the proverbial 'stranger' is what constitutes the rule, and this fear is inculcated in children and babies very early in life.

When you see that love is really a universal form of bonding among all living, a fact that is now confirmed by the existence of the quantum field, and resulting *quantum entanglement* of all living, it becomes clear that modern culture here is really upside-down, and much of the modern malaise, including large-scale depression, rape and violence are the result. As love has been shunned and denied early in Western history by patriarchy and fundamentalist religion, it has been largely replaced by *moralism* in the sense of *compulsive morality*.

The result is the destruction of *real civilization* because of the repression of the natural emotions of the child and the building of moralistic behavior structures that have gradually replaced the primary self–regulatory processes that nature has coded for the growth of all living. In other words, if we compare all of the eight black holes in the body of modern society, the denial of the love pattern through moralism has inflicted surely the most devastating damage, feedbacked by our rampant sexual violence, child neglect, missing children and large-scale murder and genocide inflicted upon *real* civilizations, and nature as a whole.

Regarding the *pleasure pattern*, we are facing a strange contradiction in the value system of modern society. On one hand society is clearly hedonistic in its overall approach toward life, and on the other it attacks and demonizes certain pleasures, almost arbitrarily. To find out what the truth is in this matter, we may investigate research on pleasure, which shows that since about the 1970s, *pleasure is scientifically recognized as a prime motivator for human behavior and decision-making*; this mechanism was intended by nature to insure procreation and healing. However, our patriarchal and puritanical past doesn't face pleasure-seeking behavior as good and healthy but

## PREFACE

used to judge it negatively. Cutting-edge research on the roots of violence has delivered clear-cut results, in that it found that *pleasure and violence are in a mutually exclusive relationship*. When we repress pleasure, we progress violence, and *vice versa*.

When we look at our data once again with this knowledge, we see that most of the pleasure modern society endorses is *ersatz pleasure*, that is, pleasure derived from material goods, while this same society keeps children from having natural pleasure with their bodies, by turning their attention toward industrially produced toys.

When we compare this situation with the situation under the rule of fundamentalist religions, the only difference is that now the repression of natural pleasure is more subtle, in that it is done, generally speaking, not through coercion but through the conditioning influence of toys, video games and television, except when it goes to adult-child sexual pleasures, where the repression is as blind, brutal and draconian as it used to be under church rule. We have seen that this, then, is the reason why our society is by far the most violent in human history, in that it has brought about murder and destruction in virtually every sphere of life.

*Self–regulation* is a pattern more obviously built into the life function than the others, in that all living processes are auto-regulatory. But symptomatically so, in a culture that is widely alienated from nature, self-regulation is also the pattern that is most commonly denied. Most people follow the precepts of some or the other authority, be it a doctor, a religious leader or guru, or a psychiatrist, to 'get on the right track,' or they follow standard moralistic or religious concepts such as 'one should not drink alcohol,' 'one should not masturbate' or 'one should only think pure thoughts.'

Less than one percent of the world population follow their own rules, set their own goals and have clear intentions. In one sample, only three percent of the followed college students were having precisely formulated goals and revised them on a regular basis. And it was only this tiny percentage, and no other criterium than persistent goal-focus, that later on produced great and unbounded success. From this perspective we can say that self-regulation properly seen means taking full responsibility for one's life.

# PREFACE

Regarding the *touch pattern*, there is quite abundant research since about the 1960s, on physical, caring touch and wellbeing. While early child psychology and even Freudian psychoanalysis were widely touch-unfriendly, mainly because of their confusion between caring touch and sexually intended touch, we saw that lacking touch produces many pathologies both within the family and on a national and global health level.

To begin with, early tactile deprivation that is produced by insufficient breastfeeding and growing career focus among mothers brings about social and individual pathologies that range from depressions over antisocial behavior until rape-centered sexuality and psychopathic behavior. On a community level, early tactile deprivation has clearly been identified as one factor in the etiology of large-scale structural and domestic violence. As a consequence of the mess-up of caring touch, shared nudity and physical closeness with so-called 'pedophilia' when care is bestowed upon children by caretakers other than their parents, and particularly when the caretaker is *male*, there are almost no more male caretakers to be found in early child care. This alone is an impoverishment so drastic that it negatively impacts upon the healthy growth of children; one factor why this is so is because children know the truth behind the lies they are told on a constant basis, and as a result they develop the same bias regarding the male as the abuser-type, and the female as the (only) caregiving-type among the sexes.

Further, no male child is able within such a distorted social paradigm to develop his maleness in a sane manner, and no female child is able to develop their heterosexual attraction strongly enough to later find a partner who is caring and homely, and thus *not* the abuser-type.

In an abuse—-centered culture, matters are messed-up and distorted to that point that the natural biogenic self-regulation has been disturbed at a level so deep that social pathologies are virtually built in this system, and simply cannot be avoided. Even if modern society doesn't follow the example of the natives, it could at least follow their own scientists who are clear-cut about the fact that touch deprivation brings about large-scale violence in any given society anywhere on the globe.

# Introduction

This introduction serves to explain certain truths about living, our universe, and human interaction that are either hardly known or seen from an ideologically conditioned point of view. Traditional Western science had no idea of the living, and of systems, it had no idea of *patterns*, it had no understanding of motion, of flow, of self-organization, of total interactive communication, and of the energy that is behind all this: the *cosmic energy field*. It was a club of adult schoolboys who secretly masturbate on public toilets and who hide behind their social masks and their doctor titles, men whose fundamental worldview is flawed by shame and denial, by sadism and the cynical and cruel attacks on the *true* scientists, men like Galileo, Leonardo, Bruno, Paracelsus, Goethe, Kepler, Mesmer, and Reich—and many others. This task would be impossible without considering the universal laws that can be known and understood by the study of, and divination with, the *I Ching*.

—See Peter Fritz Walter, The Leadership I Ching: Your Daily Companion for Practical Guidance, 3rd Edition, 2017.

#### EIGHT DYNAMIC PATTERNS OF LIVING

Studying the I Ching or the Tarot was not for me a form of *l'art pour l'art*, but it had a practical purpose: I wanted to understand the underlying *patterns of living*, and the roots of human behavior.

And indeed, my understanding of life and living was greatly enhanced through not only the knowledge about divination, but through actually practicing divination on a daily basis since more than thirty years. This deep inner experience with learning the true patterns of living and in addition my experience with helping highly problematic people to succeed in life rooted me in my new profession as a coach, and with that decision, my former career as an international lawyer clearly found an end.

My motivation for changing my professional career was the outcome of a deep reflection about the sense of life, and my mission. I understood that I am interested in human beings, and in their paths of life. The law profession, while it interested me in some way, was not satisfying my desire to help people grow, to facilitate their relationships, to help them lead *happier lives*. The I Ching supported me on my way to find my true mission, for the understanding of our behavior and motivations and those of others is

# INTRODUCTION

impossible without comprehending the underlying *cosmic patterns of living*. Human beings do not function differently than the rest of life on earth. Besides that, they are rooted in a universal scheme of cosmic interactions that is little known or totally unknown to psychologists today.

I have stressed in all my publications the importance of understanding the nature of our universe as a basically *patterned universe*, focusing on patterned intelligence, or patterned organization when we observe nature. What are patterns?

I began identifying the perennial pro-life patterns in living by firstly invalidating the fake principles that mainstream Western science declares to be the founding concepts of our universe. To put it more precisely, there was actually nothing to invalidate; I found that these alleged 'principles' were but *intellectual assumptions*, and projections, and thus simply invalid as founding principles of life. At the same time, diligent study of the I Ching and the almost daily use of it for divination distilled in me an intuitive understanding of the real and valid patterns that are inherent in all living. I therefore simply call them *patterns of living*.

Let me first of all explain why I use the term *patterns*, deciding to discontinue the use of the term *life principles*. I indeed think that here we are facing a key point that marks the essential difference between *death science* and *life science*.

A pattern is a set of things, a certain arrangement I can make out in the complex scheme of reality, and the main characteristic of this arrangement is that it forms a *relationship* of elements with each other. It is something I can observe.

A pattern can be fix or it can be changeable. It can be static or dynamic. By contrast, a principle typically is the basis of a down-hierarchy. It's a top-something in a kind of up-to-down order. It is *not* something I can *observe*. Its reality is merely intellectual: the outcome of a conclusion I draw in my rational mind *after* observing nature. A principle thus contains my observer point or my judgment about reality.

*Death science* looks at life through the glasses of principles it has set before it was going to observe. It is essentially blind and proceeds by imposing characteristics upon nature. Western science traditionally has been death science; it gained its conclusions about life by vivisecting cadavers, not by

## INTRODUCTION

observing the moving changes of living. It is, and remained, a *cadaver science* that is far removed from the changing patterns of reality.

*Life science* looks at life without any set principles or assumptions and observes the dynamic patterns or changes in the texture of life. It is a science that since its start in China, around five thousand years ago, was interested in life, and thus drew conclusions from life, and not from death. Traditional Chinese science together with most other ancient science traditions of the East is a *life science*, one branch of this very large body of science and philosophy being *Feng Shui*.

The I Ching is based upon life science, and is perhaps the highest condensation of it. Needless to add that, as such, it is non–judgmental and thus bears no moralistic judgments about human behavior. It looks at human behavior in exactly the same way it looks at all life patterns, and sees the changing nature of it before all. Fritjof Capra in his book *The Web of Life (1997)* explains the importance of pattern when he explores the meaning of *self-organization*, which is a major characteristic pattern of living systems:

> To understand the phenomenon of self–organization, we first need to understand the importance of pattern. The

> idea of a pattern of organization—a configuration of relationships characteristic of a particular system—became the explicit focus of systems thinking in cybernetics and has been a crucial concept ever since. From the systems point of view, the understanding of life begins with the understanding of pattern.
>
> —Fritjof Capra, *The Web of Life* (1997), p. 80.

In order to scientifically explore the *nature of pattern* we need to alter our basic setup of scientific investigation. Capra explains:

> In the study of structure we measure and weigh things. Patterns, however, cannot be measured or weighed; they must be mapped. To understand a pattern we must map a configuration of relationships. In other words, structure involves quantities, while pattern involves qualities. (Id., p. 81)

This really involves a radical change in our scientific thinking because traditionally Cartesian science was quantity-based and measure-oriented, while systemic science is quality-based and relationship-oriented, a truth that Capra exemplifies when looking at the properties of pattern:

> Systemic properties are properties of pattern. What is destroyed when a living organism is dissected is its pattern. The components are still there, but the configuration of relationships among them—the pattern—is destroyed, and thus the organism dies. (Id.)

INTRODUCTION

The next important point to understand how nature *thinks* is the cell's metabolism, the network that serves recycling. Capra succinctly elaborates in his book *The Hidden Connections (2002)*:

> When we take a closer look at the processes of metabolism, we notice that they form a chemical network. This is another fundamental feature of life. As ecosystems are understood in terms of food webs (networks of organisms), so organisms are viewed as networks of cells, organs and organ systems, and cells as networks of molecules. One of the key insights of the systems approach has been the realization that the network is a pattern that is common to all life. Wherever we see life, we see networks. (…) The metabolic network of a cell involves very special dynamics that differ strikingly from the cell's nonliving environment. Taking in nutrients from the outside world, the cell sustains itself by means of a network of chemical reactions that take place inside the boundary and produce all of the cell's components, including those of the boundary itself.
>
> —Fritjof Capra, The Hidden Connections (2002), p. 9.

But the most revolutionary outcome of the systems view is that our usual habit of dissecting parts of a whole for further scrutiny and scientific investigation *does not work* with living systems. Why is this so? Capra pursues in *The Web of Life (1997)*:

> Ultimately—as quantum physics showed so dramatically—there are no parts at all. What we call a part if merely a

pattern in an inseparable web of relationships. Therefore the shift from the parts to the whole can also be seen as a shift from objects to relationships.

—See Fritjof Capra, The Web of Life (1997), p. 37.

My hypothesis is that Western culture has *never* until now applied the *Eight Dynamic Patterns of Living* and that it *therefore* is at the *border of chaos, destruction or another kind of worldwide catastrophe;* I allege that this culture is suffering from a schizoid mindset, the perversion of love into sadistic hate, rampant violence, the impudent slaughtering of ethnic and cultural minorities, famines that could easily be avoided, and generally a total lack of genuine spirituality which, by itself, already makes for a large part of depression and psychosomatic disorders.

What I say is that the *Eight Dynamic Patterns of Living* have been respected and applied by all major tribal cultures including the North American Indians, and that *therefore* they have lived peacefully. With 'peacefully' I do not mean an artificial *Western* peace concept which is complete nonsense as it is stuck and rigid, but a *dynamic* peace continuum that includes little fights and small wars as required by the

## INTRODUCTION

dynamics of *yin* and *yang*, but that is so balanced that it will never trigger a major and global destruction. The fact that Western culture has triggered this destruction in all possible ways, economically, socially, health-wise, militarily and ecologically shows that the *continuum balance* that the *Eight Patterns* give is completely lacking in Western philosophy, science, military policy, diplomacy, politics and strategy. Western culture has brought about what Wilhelm Reich called *the emotional plague*, symbolized by the atomic bomb's mushroom.

The *Eight Patterns of Living* could be taken as a guide concept for being implemented in a new kind of lifestyle to be worked out as part of our presently evolving postindustrial global culture. That is the basic idea.

Besides, I think that the *Eight Patterns of Living* are tremendously useful as a base layer for establishing the principles of a new peaceful society, instead of beginning with Adam and Eve and go time and again through all anthropological material. I have actually done this and found that there is no novelty any more in this. The *Eight Patterns* cover all spheres of life and living.

# The Eight Patterns (Overview)

I shall first give an overview over the eight patterns, as little paragraphs, and then discuss each of the patterns in the next chapters. This overview is of course incomplete; it is destined to give you an approximation of what each pattern is about.

## 1) Autonomy

All peaceful tribal societies have in common that they grant their children an utmost of autonomy. In *dominator cultures*, that today represent the bulk of large and typically industrialized societies worldwide, the lacking autonomy of the consumer child is a truly pathological phenomenon that often takes the form of co-dependence, which I call *symbiotoholism* or emotional abuse and in general the unhealthy fusional clinging of members of the family, or as collective fusion through the identification with groups, organizations and ideologies. In fact, observing the growth processes in nature, we can see that autonomy is something built in all living, and as such takes part in all growth.

In order to realize our personal identity and become whole human beings, we have to be able, still in childhood, to form an original personal identity. This is however impossible if we are reared by narcissistic parents, those namely that are *indifferent* to the unique person of the child they have brought to life.

## 2) Ecstasy

All peaceful tribal societies have in common that they have a strong *ecstasy pattern* built in their lifestyle which makes them once in a while enjoy group events where the usual rules of conduct are more or less set aside. Usually, these events are characterized by magic rituals, the consumption of mind-altering *entheogens*, that is, psychedelics, and the partial or total disregard for the incest taboo or other sexual taboos.

This principle was wide-spread even among major civilizations; still some decades ago, during the

*Carnival in Rio*, it was not uncommon that sexual incest was practiced between parents and children. It is also quite probable that intergenerational sex, while practiced in very few aboriginal cultures, is allowed on a larger scale also in less permissive cultures during *ritual events* that serve to liberate and cultivate individual and group ecstasy.

> —Interestingly, neither Bronislaw Malinowski nor Margaret Mead found sexual paraphilias present in Melanesia's Trobriand culture where children enjoy utmost emotional and sexual freedom.

## 3) Energy

Life is energy! This is recognized as a vital life pattern in all non-Western societies, and thus the overwhelming part of the world. Oriental cultures were historically the most wistful in recognizing and applying energy patterns for healing, good fortune and positive relationships.

The Chinese science of *Feng Shui* is perhaps the oldest distillation of this holistic knowledge into something we today call a *science* while traditionally Orientals tend to speak rather of *philosophy* or of *religion* when they talk about the perennial science of the bioenergy.

However, even in the West, alternative scientists from Paracelsus to Reich have acknowledged the existence of the *bioenergetic functionality* not only of the human organism, but also of the weather, the atmosphere and the cosmos as a whole.

While in substance these researchers observed basically the same phenomena, the way they termed the cosmic life energy varied. Paracelsus spoke of *vis vitalis*, Swedenborg of *spirit energy*, Mesmer of *animal magnetism* and Reich of *orgone*. And since millennia this same energy was called *ch'i* by the Chinese, *ki* by the Japanese, *prana* in India and *mana* with the Kahunas from Hawaii and the Cherokee natives of North America.

Furthermore, parapsychologists universally agree that the motor of all psychic phenomena is to be found in our bioplasmatic and egg-shaped *aura*, an energy body of lesser density that we carry around our physical body and which can be seen as an extension of our bioplasmatic energy, as it is composed of the same bioenergetic charge that we find in the bioplasm.

Emotions are energetic, streaming currents that are a direct outflow of the cell's bioplasma. I speak about emotional flow or, within my bioenergy research, of *emonic currents*. These streamings have their seat not in the brain, as modern psychology wrongly believes, but in the bioplasm and in the aura, our subtle etheric energy body.

## 4) Language

Psychoanalysis has revealed the importance of language as a condition for the sublimation of instincts. Furthermore, peace researchers found that a lack of language and thus of communication is at the basis of all forms of violence, inner and outer. This insight has not only psychological but also political consequences. For it clearly indicates that only free speech and democracy, both within the family and the nation, can ensure maintaining peace and regulating our natural instincts and desires, so that they do not become asocial and violent through denial. To everyone who says that we have democracy and yet are a violent society, I reply that we do not have true democracy and never had. For violence only comes up when verbal communication is impaired, and the one major reason why communication is impaired about vital issues is *shame*. When we feel ashamed about certain vital events in life, such as sexuality, we do not freely communicate about these issues, because we are blocked or inhibited by the nagging feeling of toxic shame that comes up every time we tackle the subject.

Lack of communication leads to violence; where the mouth is defended to talk, the body takes over the role of the mouth—and the fist talks! We all know this from history and from private experience, and yet there is little general conscience in our society about the almost sacred importance of dialogue, of communication, not only outside, in relationships with others, but first of all inside, in the relationship with ourselves.

Our large civilizations do very little to integrate the wisdom of language because they are hardly conscious of the power of the word. Tribal cultures, however, are wistful in this respect and generally dispose of an array of rituals that serve precisely the purpose of what in our civilizations we do within a psychotherapy: putting words on things, events and feelings.

## THE EIGHT PATTERNS (OVERVIEW)

# 5) LOVE

All peaceful tribal societies have in common that they follow the *love pattern* and not, as most of the larger nations, the morality principle. The present state of violence within and between our larger civilizations, especially those with *high morality* is in my view the result of despising the love principle and the widespread use, also and especially in politics, of *moralism*. In other words, it is the disregard of one of nature's highest principles, the principle of *biogenic self–regulation*, that brought about the present state of violence and the lack of love and true care among most of the peoples of the earth. It is the hypocrite manner of preaching peace and democracy by our false and opportunistic leaders while they treat natural love like the Biblical serpent.

Wilhelm Reich, in his extensive work on the psychological roots of fascism found that it is the repression of our natural emotions, first of all by prohibiting *our young generations the natural acting-out of their love desires* that brought us at the border of the present abyss of fundamentalism, persecution, slaughter, genocide, war, civil war and worldwide terrorism.

—See Wilhelm Reich, The Mass Psychology of Fascism (1933).

## 6) Pleasure

All peaceful tribal societies have in common that they acknowledge the *pleasure pattern*, for example in the way they educate their children. In planning the child's future, what counts is *not the father's job*, that is the typical dominator position, but the natural inclination and interest of the *child* for their later profession. By doing this, instead of projecting upon children their parents' wishes and desires, *education ensures* that every generation does in most of their time what they really are gifted for. The result is both a high level of skill and motivation for profession and career.

It is not surprising that now also in modern nations the *pleasure function* begins to be seen as the main motivating factor for a person's advancement in life.

Suffices to read the biographies of great and successful men and women to see that all achievement is a result of desire and persistent acting upon desire and that there is no better catalyzing agent than biological self-regulation that is based upon pleasure.

## 7) Self–Regulation

All peaceful tribal societies have in common that they follow patterns of self–regulation or *permissiveness* in the education of their small children, and consequently restrain from inflicting violence in form of physical punishment upon them. The most peaceful of those tribal nations, the *Trobriands* of Papua New Guinea are completely permissive as to children's sex play and early mating games.

## 8) Touch

All peaceful tribal societies have in common that they are conscious about the *touch pattern* and care for maintaining free body touch among family members, nudity, and abundant tactile nutrition for infants and small children in the form of baby massage, carrying the baby on the body, and sleeping naked with children. In dominator cultures, life-denying pediatricians were turning down parents' desire for fondling their children and co-sleeping naked with them.

Now we slowly begin to see the macabre results of the *deprivation of tactile stimulation* in infants. Psychosomatic medicine more and more reveals that our immunity against viruses depends on touch and that lacking touch, especially in childhood, leads to more or less acute *immune deficiency* and as a result to higher vulnerability for certain *lifestyle diseases*.

Furthermore, cross-cultural research has clearly shown that *early tactile deprivation* is one of the major inducing factors for the plague of personal, domestic and structural violence in any given society.

# The Autonomy Pattern

*Autonomy* is fundamental for each and every being-in-growth. Without autonomy, there is fusion, symbiosis and dependence. While for certain organisms, such as the human newborn, symbiosis for a certain time is a biological necessity, this symbiosis is time-bound and should gradually give rise to autonomy.

While natural symbiosis is needed for the first eighteen months of the newborn, it should gradually come to an end after that period. Unfortunately, postmodern international culture is *more or less completely dysfunctional* regarding this primal movement from fusion to autonomy that should take place, dynamically, in the growth process of the human baby.

What happens is that the necessary biological symbiosis with the mother, *eighteen months from birth*, is neglected for various reasons; babies suffer from a more or less stringent tactile deprivation that will leave scars for their whole lives.

In order to compensate for the lack of care bestowed upon the infant, as a guilt-reaction and for various other reasons, the post-symbiosis condition is not better for the child: instead of growing into autonomy most children in postmodern international culture grow into codependence with their parents and caretakers; instead of building a gradually larger extent of autonomy, parents tend to gradually entangle their children in a tight net of stiffening dependencies; this form of *emotional vampirism* is so rampant especially in modern Western societies that I have termed it *emotional incest*.

I further argue that the present defamation and persecution of *affectionate and nurturing erotic love between children and adults* and its confusion with child-endangering sexual sadism have their origin in shame and guilt Western society is suffering from because of its deprivatory and dysfunctional childrearing paradigm that endorses and purports emotional and, in a hidden way, sexual incest by holding children dependent, helpless and infantile as long as possible so as to *compensate for the crippled*

## THE AUTONOMY PATTERN

*bioenergetic structure of their parents and caretakers.* This is how an ever new generation of emotional and sexual cripples will raise an ever next generation of dysfunctional water–headed babies that are going to live with a perverted bioenergetic base structure.

Such a situation is *shameful* for a society based upon *egalitarian principles* and that pretends to respect the *person of the child;* however, this reality is veiled, because shame tends to bring about defensive and projective reactions. The projections are clearly to be seen in the fact that the majority of psychologists, psychiatrists, physicians, psychoanalysts or psychohistorians, without further information assume that *pedophilia* was something like a *metaphorically incestuous behavior,* and that it *therefore* was offending society's written or unwritten behavior code. This concern, that has been voiced with particular stress by Lloyd deMause, founder of *Psychohistory,* appears to be rather of an ideological nature and fails to stand against the very principles it is founded upon.

—See Lloyd DeMause, Foundations of Psychohistory (1982).

Psychohistory is not a morality codex, but a science that regards world history under psychoanalytical perspective, applying to historical events and human motivation the psychoanalytical method as it was developed mainly by Sigmund Freud. Yet Lloyd DeMause steps out of scientific objectiveness when he associates pedophile sexuality with incest, concluding that pedophilia *was after all* a *prolongation of incest* beyond the borders of the family. I hold against this argument that it is in itself *incestuous*.

In order to demonstrate what is right and what is wrong here, I have to dig a little deeper and get back to the foundations, not of psychohistory, but of psychoanalysis. In fact, Freudian psychoanalysis affirms the sexuality of the child, and newer research has shown that even the fetus is sexual in an auto-erotic manner.

While it is true that Freud took a distance to his student and collaborator Wilhelm Reich because the latter began to fight politically for children's sexual freedom, Freud rejected Reich not because of a divergence in scientific perspective. Their split was a mere cultural controversy, or, as we would say today, a matter of *political correctness*. Freud was convinced of the child's abundant sexual life, but he thought, as

## THE AUTONOMY PATTERN

he literally replied to Reich, that *culture primes* and that we had to respect Western society's fundamental denial of children's sexual freedom.

As a result, Freud, and after him the overwhelming part of psychoanalysts, simply blinded out from their professional regard *any sexual activity of the child outside of the family.* In fact, when you read psychoanalytic writings, you are overwhelmed by the extreme focus of these people upon *incestuous wishes* of the child or both the child and the parents. The Socratic error here is to assume that this view was *scientific* in any way, while it is truly the consequence of a cultural, and in addition a professional *bias*. The cultural bias is the fact that in patriarchal societies, *natural* sexuality is forbidden for children with the result that *unnatural* sexuality is brought about, mainly in the form of rape-centered pornography, sadomasochism and violent child rape, abduction and murder. The professional bias is the fact that psychoanalysts typically deal with neurotic, and not with sane people.

The final and quite far-reaching results of this fundamental position of mainstream psychoanalysis are the following. The regard upon sexuality is distorted in that incest and incestuous wishes are viewed upon with an exaggerated and unnatural focus that veils the fact that the human being, if raised freely, naturally projects sexual wishes outside of the family. The factual *oedipal touch* of the modern nuclear family and the really widespread problem of incestuous wishes, and factual emotional and sexual incest, is the result of patriarchy's denial of *child-child sex* outside of the family; anthropological research within tribal cultures that give their children full emotional and sexual freedom for copulating with other children corroborates that in these societies incest is absolutely non-existent; interestingly enough, what also is practically non-existent in these cultures is violent crime and sexual dysfunctions, as well as homosexuality and pedophilia.

As modern society says violence is good, and sex is bad, it by the same token says incest is the rule for the child, while sex outside of the family is the exception as it is invariably considered as criminal and *allegedly damages and traumatizes the child*. This is how patriarchy has put nature upside down: it focuses upon the sick and dysfunctional and disregards the plain, healthy and natural.

# THE AUTONOMY PATTERN

Lloyd deMause and many of his colleagues in faith seem to suffer from a blind spot as they virtually *project* incest outside of the family, not seeing that they confuse the rule with the exception. Suffices to read, for example, Françoise Dolto, who was a mainstream Freudian psychoanalyst with an outspoken Christian orientation. Dolto was in France and later in life internationally renowned as child therapist, and she expressed in her books the view that the healthy child naturally wishes to find love mates of same or different age outside of the family.

—See Peter Fritz Walter, Françoise Dolto and Child Psychoanalysis, Great Minds Series, Vol. 4 (2015/2017).

Dolto when asked if she did not agree with Freud that every child had incestuous wishes toward his or her parents, replied this to young psychoanalysts during one of her workshops on child psychoanalysis:

> You as psychoanalysts have to deal, in your daily practice, with neurotic children. Of course, neurotic children are incestuously fixated, because the very etiology of neurosis, as we know since Freud, is sexual. So, with this bias in your mind, you wrongly assume that the same was true for the healthy child.
>
> —Françoise Dolto, Séminaire de Psychanalyse d'Enfants, Tome 2, (1985), p. 21 (translation mine).

What is true for sexuality is equally true for *autonomy*. A naturally sexual child is typically *more independent and more autonomous* than a neurotic and incestuously fixated child. The frequently observed *clinging behavior* of modern city children, their helpless, infantile and irresponsible behavior, even as late as when approaching puberty, their immaturity in handling sharp or fragile objects such as knives or glasses show well their incestuous fixation, their neurotic blockage and codependent entanglement with their parents, and the early psychosexual damage a life-denying and pleasure-hostile education inflicted upon them.

There is a natural striving for autonomy built into every growing life. For example, a child of three years of age needs more autonomy than a child of fifteen months of age; a toddler of eighteen months needs more autonomy than a baby of five months. Many parents ignore that babies, toddlers and pre-schoolers, already before reaching the age of primary school, need to develop autonomy. Many adults believe that children grew through magic shifts, like the one from babyhood to childhood, from childhood to youth and from youth to adulthood. The

## THE AUTONOMY PATTERN

first shift is believed to take place around seven years of age, the next one around twelve years of age and the final one around eighteen years of age.

Sorry, but this is really myths. These shifts don't exist in real life as all growth is gradual and smooth. This is why all education should be gradual and smooth. While it is a good thing to have certain initiation rites or ceremonies that mark important steps in the growth of children, these rites are what they are: mark stones that border an otherwise seamless road. I arrive at a mark stone, I see the mark stone, I touch the mark stone, I pass the mark stone, I remember the mark stone. My passing the mark stone is gradual, and smooth in time, and the mark stone itself is of lesser importance than my passing it. What is important is that I constantly grow, that I remain *moving*. We learn the basic *movement into autonomy* during our *first year of life*, and not later on during adolescence or when we allegedly *turn into* that magic world of adulthood.

I do not want to belittle the important changes that take place in the life of adolescents, and their sometimes passionate focus upon getting more autonomy, nor do I belittle the marking shift from adolescence into final adulthood. But often we observe that especially those adolescents who have rather repressive and possessive parents get onto the obnoxious track and really push it through for every millimeter of increased autonomy. There is a logic in every behavior and adolescents who put high stress on autonomy *have a reason* to do so. The reason is rooted in much earlier years, in the years of babyhood.

*There is no alternative to autonomy.* To make down or belittle children's need for autonomy is to open the door to emotional and power abuse, and *large scale*. This form of child-abuse is not perpetrated by the proverbial *stranger*, but by *mothers*, first of all mothers, and more and more also by professionals who are working for, affiliated with, or sponsored by the international *child protection* industry.

# The Ecstasy Pattern

What is ecstasy—is it merely a pleasure boost? If it was that, this pattern would have to be annexed to *Pattern Six: Pleasure* and a separate pattern would not be needed.

To avoid a lifeless theoretical discussion on this point, let us proceed empirically and *observe*.

Does a sex orgy bring about ecstasy? I have shortly mentioned in the overview over the eight dynamic patterns above that in fact there were in the past taboo-free events in many cultures, usually one or two days per year where incest, even in direct line, was given a free license. It is historically documented that even in major cultures such as Italy and Brazil this custom reigned still in the Renaissance during the *Carnival in Venice* and the *Carnival in Rio*.

Many tribal cultures have similar rites, but after deeper research into shamanism and having myself gone through a psychedelic experience, I came to the conclusion that ecstasy in the true sense cannot be triggered merely by fulfilling one's sexual wishes, be it wishes that normally remain repressed in our culture. We still are with the pleasure function when we think of orgiastic events, intimately or in public, in group sex orgies, we still have a simple sexual satisfaction, an orgasm, at the end. Okay, it may be enhanced through the exotic nature of sex, with children or even one's own children, for example, or even in front of others who do the same, but still, ecstasy as I have found it present in, for example, the native *Shuar* culture in Ecuador, is something entirely different than high sexual fulfillment. While I admit that sexual fulfillment is very important for a healthy life and that we should have as little as possible sexual restrictions, *this is not all there is* when we look over the fence of mainstream conditioning.

Ecstasy induced by psychedelics is *not* related to the pleasure function, while in some trips there are actual sexual encounters and strong and satisfying sexual feelings, but this is rather the exception. What I found characterizes the truly ecstatic experience is *religion*, religion in its original and pure sense: a backlink established to our soul and our profound holistic wisdom. It's generally not really a pleasurable experience: it's in most cases an awe-inspiring experience of deep wonder and bliss. And one goes out humbled and wistful as a small child.

Sexual pleasure and orgiastic satisfaction, certainly important as such for our health cannot really be compared with the wonder and the ecstasy that are the result of a psychedelic experience where *set and setting* are correctly chosen.

### Ralph Metzner

It is widely accepted in the field that set and setting are the most important determinants of experiences with psychedelics, while the drug plays the role of a catalyst or trigger. This is in contrast to the psychiatric or psychoactive drugs, including stimulants, depressants and narcotics, where the pharmacological action seems paramount, and set and setting play a minor role.

—See Ralph Metzner in Ralph Metzner (Ed.), Ayahuasca, Human Consciousness and the Spirits of Nature (1999), p. 24.

It seems that the opening of mind is a larger experience generally than the fulfillment of desire in the form of orgiastic pleasure. In many tribal cultures, in fact, both experiences are linked to each other during festivities, and perhaps there is something like one feedbacking upon the other and enhancing the other? Further research is needed here.

All the literature I found relating to ecstasy was exclusively dealing with entheogen-induced ecstasy, or ecstasy as a religious, not a sexual, experience.

Therefore, let me report this research more in detail instead of putting up assumptions about a greater vision of ecstasy that I cannot for the moment backup with scientific data. Terence McKenna, when asked by Jay Levin to define *shamanism*, replied:

> Shamanism is use of the archaic techniques of ecstasy that were developed independent of any religious philosophy—the empirically validated, experientially operable techniques that produce ecstasy. Ecstasy is the contemplation of wholeness. That's why when you experience ecstasy—when you contemplate wholeness—you come down remade in terms of the political and social arena because you have seen the larger picture.
>
> —Terence McKenna, The Archaic Revival (1992), p. 13.

This is why I engaged my shamanism and entheogens research in the first place: it was for elucidating the *Ecstasy Pattern* within the research project for the present production. But what *is* ecstasy, then? Terence McKenna, in his book *The Archaic Revival (1992)*, explains at p. 144: 'Ecstatic is a word unnecessary to define except operationally: an ecstatic experience is one that one wishes to have

## THE ECSTASY PATTERN

over and over again.' McKenna is right in not complicating with what anyway can hardly be verbalized. Ecstatic joy is different from sensual satisfaction because it is a *learning experience*. We learn each time about ourselves and the universe. That is why ecstatic experiences are often also called *mind-opening* experiences. But there is more. When you remember your childhood you may recall a form of excitement that was not related to sexuality, but to some form of wonder, a joy that you could not explain where it was coming from and what it was induced by. I have called it *feeling good without reason*.

You may remember that as a child at times you felt like flying right to heaven, so happy and excited you were, and most of the time without any apparent reason. Well, I succeeded in reawakening this basic innocence in myself and thus I do experience moments of full and unhampered happiness and ecstasy when they come; and they come spontaneously, without being asked for and without being triggered by any drug.

But I know that not many adults have kept their childhood innocence or have achieved reawakening it during adulthood.

Now, let me address a fundamental argument that most people come up with more or less spontaneously regarding the psychedelic experience: they tend to argue that drugs will render you *addictive*. This is true for certain drugs, but not for all drugs. It is true for sugar, for alcohol, for coffee, for tobacco, for most tranquilizers, for some medical drugs, for some of the hard core drugs such as heroin, but it is *not* true for cannabis and what we call psychedelics or entheogens.

Shamanism researchers like Michael Harner, Adam Gottlieb, Ralph Metzner or the McKenna brothers have repeated often in their publications that psychedelics do not render addictive; regarding synthetic drugs, we should remember that once LSD was a scientific substance, used in psychotherapy, and that the one who discovered it, the Swiss chemist Albert Hofmann, asserted that LSD was not rendering addictive in any way. Why the drug ultimately was put on the index of forbidden drugs has nothing to do with a dependency problem, but simply and clearly has political reasons. Terence McKenna explains:

> The solution of much of modern malaise, including chemical dependencies and repressed psychoses and neuroses, is direct exposure to the authentic dimensions

of risk represented by the experience of psychedelic plants. The pro–psychedelic plant position is clearly an anti–drug position. Drug dependencies are the result of habitual, unexamined, and obsessive behavior; these are precisely the tendencies in our psychological makeup that the psychedelics mitigate. The plant hallucinogens dissolve habits and hold motivations up to inspection by a wider, less egocentric, and more grounded point of view within the individual. (Id., p. 219)

Most experts on shamanism agree with Mircea Eliade who says that the role of the shaman in shamanistic cultures is *to be a manipulator of the sacred, whose main function is to induce ecstasy in a society where ecstasy is the prime religious experience.*

—Dennis McKenna & Terence McKenna, The Invisible Landscape (1993).

The subtitle of Mircea Eliade's study *Shamanism: Archaic Techniques of Ecstasy (1989)*, which is the classic of all classics on shamanism clearly suggests that shamanism is *primarily a system of techniques destined to bring about and maintain ecstasy*, individually and collectively, as a repeated experience so vital for the tribe and tribal life, and for peaceful social togetherness within the tribe and between tribes, that its importance cannot be overestimated.

Why should ecstasy, then, be so important for peaceful living, for social relations, for peace, for a wistful attitude in living? What is so special in ecstasy, and what is different in ecstasy compared with the satisfaction of desire, or sensual and sexual pleasure? Terence McKenna has answered this question when he said that ecstasy is the *contemplation of wholeness*. Thus, in simple terms, when we talk about ecstasy, we talk about religion, not religion in its perverted function as a system of indoctrination or morality catalogue, but religion in its original sense as a *gnosis*, a gain of knowledge about life, an instantaneous holistic revelation about the *deeper meaning* of our lives, a regard that implies the gain of *direct intuitive knowledge* about the sense of living, the meaning of our lives, and the essential nature of *all–that–is*.

In this sense, there is a synchronistic link between shamanism and the experience of ecstasy as the contemplation of wholeness, on one hand, and paranormal realities such as the fairy world, the religious visions and miracles experienced by sages and saints, and generally, the discovery of *soul* in our daily life, on the other.

## THE ECSTASY PATTERN

—See, for example, W.Y. Evans–Wentz, The Fairy Faith in Celtic Countries (1911) and Dora van Gelder, The Real World of Fairies: A First-Person Account (1999).

Thomas Moore, in his bestselling book *Care of the Soul (1994)*, gives practical guidelines about how to bring about authentic ecstasy within our own culture; as such the book is of unprecedented value, a precious gift that goes beyond all research reports about shamanism because it gives us, within our own culture, ways of bringing about ecstasy in daily life and in our multiple relationships with self, other and the whole universe.

A similarly precious guide for realizing the authenticity of our own soul reality, while coming from a quite different point of departure, is Karen Kingston's book *Creating Sacred Space With Feng Shui (1997)*. Karen Kingston's unique talent is an inborn and absolutely stunning natural intuitive sensibility for the higher dimensions of existence, for all that is invisible to our eyes and undetectable by our five senses. The novice reader may be astonished about the authority that this text reveals and the power of the author's approach to Feng Shui that is the pragmatic and direct approach of an experienced practitioner. This book is not simply one of those poetic writings that elaborate a *magic* view of life. It is that also, but it is much more. Behind the beautiful appearance and the refined language is hidden a *hard core* manual that is truly scientific—in the sense of a universal and holistic form of science.

Of course, the representatives of reductionist modern science would question most of Karen Kingston's scientific concepts. But this argument is true for almost all publications about Feng Shui, which is not a collection of ancient myths, or superstition, but a real science.

> —See Peter Fritz Walter, Basics of Feng Shui: The Art and Science of Sensing the Energies, Scholarly Articles, Vol. 11 (2015/2017).

What Karen Kingston does is precisely to go beyond the limits of a clockwork science that is based upon today admitted wrong premises about life. To call Karen Kingston's approach to life *animistic*, an argument that has been put forth also against most of Goethe's scientific writings of, and first of all against his *Color Theory*, would disregard the deep and intuitive truth that is at the basis of this holistic life philosophy.

# THE ECSTASY PATTERN

—See Johann Wolfgang von Goethe, The Theory of Colors (1970), and Frederick Burwick, The Damnation of Newton (1986).

Karen Kingston who is married with a Balinese and lives several months every year in Bali, gives pertinent information in her book about the ways that Balinese use Feng Shui or *Space Clearing*.

There are in Bali actually two levels of handling spiritual wisdom, a professional level—if I may say so—and a popular or intuitive level. The professional level is since many generations in the hands of the first caste, and especially the *Pedandas*. Here, we encounter a highly sophisticated and informed way of handling spiritual information that is so complex and so deep that most Westerners would only shake their heads when they heard about it. On the other hand there are the people in the street who, in Bali, it seems, are also wiser as anywhere else in the world. For they, too, have this knowledge, only in a more intuitive and less literary form.

Having myself lived and worked Bali for several years, I understand Karen Kingston's natural affinity with Bali and the Balinese. I could not imagine where else somebody like her could live. It all sounds like a miracle but I am convinced that we will fully understand it once we know more about the complex influences that sound and vibrations have on our aura, on all our seven etheric bodies.

From what we learn from such an experience, it appears that neither intellectual brilliance nor extraordinary talent or knowledge, nor else a specific sense for paranormal realities, but simply total *surrender* is needed to experience the transformation entheogens bring about. McKenna states in his book *Food of the Gods (1992)*:

> Shamanic ecstasy is an act of surrender that authenticates both the individual self and that which is surrendered to, the mystery of being. Because our maps of reality are determined by our present circumstances, we tend to lose awareness of the larger patterns of time and space.
>
> —Terence McKenna, Food of the Gods (1992), p. 7.

That the *ecstasy pattern* has never been integrated in patriarchal civilizations is kind of logical. This denial is the inevitable outcome of the power structures in a culture that knows nothing else than dominating and exploiting nature instead of understanding it. The knowledge taboo is inherent in patriarchal or dominator society and it finds its explanation simply in the fact that a fundamentally undemocratic oligarchy always focuses upon manipulation, not information, indoctrination, not natural knowledge to keep the masses at stake and itself at power.

It has been established since the beginnings of the 20$^{th}$ century, foremost by Edmond Bordeaux-Szekely, and his discovery, in the Vatican library, of the *Dead Sea Scrolls*, the original gnosis of the Essenes, in which sordid ways the Church falsified the teaching of Jesus of Nazareth and the whole of the text of the Bible.

Do we need further proof for the fact that, as a culture, we have gone astray from the original and immediate knowledge that direct perception conveys? Direct perception is the original mode of knowledge gathering used by most tribal cultures that live in accordance with the wisdom of nature. In these cultures, most use entheogens such as *Ayahuasca* to get in touch with the perennial wisdom of plants and mushrooms, beings by far older than our human race.

Of course, in our dominator societies, this knowledge has been largely forgotten because of the knowledge repression terror of the Church's Inquisition for more than one millennium.

> —Schultes, Hofmann and Rätsch report that mushrooms are among the most archaic forms of living on earth and have therefore assimilated an infinite amount of knowledge about evolution on earth that is not per se accessible to a human being without access to entheogenic substances. See Richard Evans Schultes, Albert Hofmann, Christian Rätsch, Plants of the Gods (1992).

Yet, direct perception is the angular stone of J. Krishnamurti's unique teaching that revived this ancient knowledge for our times, and it has been found by famous think tanks such as Edward de Bono and learning innovators such as Dr. Georgi Lozanov, originator of *Superlearning*, that direct perception is the *primary mode of functioning* of our brain; it was also asserted that it is our original, unspoiled, and highly effective mode of learning.

> —See Sheila Ostrander & Lynn Schroeder, Superlearning 2000 (1994).

After the analysis, let us have a look at how we can *integrate* our lost Gaia knowledge if we want to. The late Terence McKenna, eminent expert on the matter, was rather skeptical because access to entheogens has become tiresome and difficult with modern

## THE ECSTASY PATTERN

society's problematic stance on mind–altering substances:

> The impact of hallucinogens in the diet has been more than psychological; hallucinogenic plants may have been the catalysts for everything about us that distinguishes us from other higher primates, for all the mental functions that we associate with humanness. Our society more than others will find this theory difficult to accept, because we have made pharmacologically obtained ecstasy a taboo. Like sexuality, altered states of consciousness are taboo because they are consciously or unconsciously sensed to be entwined with the mysteries of our origin—with where we came from and how we got to be the way we are. Such experiences dissolve boundaries and threaten the order of the reigning patriarchy and the domination of society by the unreflecting expression of ego.
>
> —Terence McKenna, Food of the Gods (1992), p. 52.

On the other hand, international culture has transformed the world into a village and we can travel to cultures with a still largely shamanic world view, at least in their more remote parts, such as, for example Ecuador. In fact, in 2004, invited by a Polish businessman married with a *Shuar* native, I went to Misahualli, Ecuador for experiencing the ritual ceremony of taking the Ayahuasca brew, in a context where set and setting were carefully chosen and not left to hazard or tourism–friendly amateurism.

After this experience with Ayahuasca, I am convinced that we can directly access nature's original wisdom and receive guidance from its universal intelligence, as well as benefit from a power of deconditioning that has no parallel in human history and experience.

# The Energy Pattern

All native cultures have an innate sense for the energy nature of life, all life, not only human life.

> —My study of the age-old Chinese science of Feng Shui from 1998 has revealed me clearly that already as a child I had the faculty of directly sensing the bioenergetic charge emanating from people and being present in certain locations. I still have this faculty today.

This fact has long been veiled because of the metaphoric language used by most tribal cultures. For example the amazing knowledge that the *Dogon* in Mali have about the cosmos and even details about certain stars, such as Sirius, knowledge that in our culture only astronomers have, is all coded in a beautiful metaphoric language.

> —See, for example, Robert Temple, The Sirius Mystery (1998), Part One, pp. 53 ff.

The same has been observed by Terence McKenna as to the terminology used by tribal peoples to describe phenomena pertaining to the human energy field. In *Archaic Revival (1992)*, and with regard to the bioenergetic charge contained in plant substances used for religious purposes, McKenna writes:

> They are the true phenomenologists of this world; they know plant chemistry, yet they call these energy fields *spirits*.
>
> —Terence McKenna, The Archaic Revival (1992), p. 45.

It has often been objected to the amazing knowledge of tribal populations that shamanic reality was a pre-psychotic or *primitive* worldview that only peoples with an archaic mindset could uphold, but that, by contrast, the Western citizen had a far higher developed consciousness.

This kind of ignorant judgments that used to be uttered by anthropologists and psychologists suffering from the usual Christian and colonial conditioning, fortunately have largely been silenced by advanced and unbiased research on shamanism and parallel realities; what strongly corroborates the evidence of plants being *universal radios* of life-related knowledge is the fact that, contrary to prejudice, Western people *who have had exposure to plant teachers* react to this experience in much the same way as natives do, and report pretty much the

## THE ENERGY PATTERN

same phenomena as natives report and reported over millennia. What namely recurs in these *visions* or *insightful journeys*, or *contemplations of wholeness* or however one may want to name these explorations in the plant realm, is the fact that nature is basically coded in *energy patterns*, these patterns being observable, recognizable and subject to conscious manipulation for beneficial or for detrimental purposes.

Black magic, sorcery, is explainable as a conscious manipulation of energy patterns for the sake of hurting others, and from this point of view and after scrutiny it becomes evident that sorcery is a powerful school of wisdom, as has been shown, *inter alia*, by Carlos Castaneda's explorative journeys with a *Yaqui* sorcerer from Mexico.

—Carlos Castaneda, The Teachings of Don Juan (1985), Journey to Ixtlan (1991), Tales of Power (1991), The Second Ring of Power (1991).

I will now quote three references from Ralph Metzner's sampler *Ayahuasca, Human Consciousness and the Spirits of Nature (1999)*.

Kate S., a woman artist in her forties, reports:

I had the thought that the reason certain cultural or ethnic art forms appear is because of the *planetary energy* in the location of the origin of that form, and that the music and the art were intricately connected and reflective of the energy of the planetary location of their origin and the energies which exist there. (Id., p. 72)

I.M. Lovetree, an educator in his fifties, states:

After the ayahuasca sessions, I feel cleansed within, throughout and all about. I have a sense of having been healed at all levels, especially the physical. The ayahuasca medicine seems to have a special affinity for the gastrointestinal system: it snakes its way through the body, seeking out and eliminating obstructions to *life energy flow*. I sometimes think of it as a form of kundalini, a Liquid Plum'r for the soul. For cleansing and healing, for reconnecting with the vegetable kingdom, ayahuasca is definitely my medicine of choice. (Id., p. 123)

Ralph Metzner, in concluding the reader, summarizes:

The fundamental reality of the universe is a continuum, a *unitive field or fabric, of both energy and consciousness*, that is beyond time, space and all forms, and yet somehow mysteriously within them, simultaneously transcendent and immanent. In traditional Asian religions, this unitive field is variously referred to as Tao, or Atman–Brahman or Tantra (the 'web' or 'fabric' or the 'jeweled net of Indra'). Some Native North Americans refer to it as Wakan–Tanka, the all–pervading Creator Spirit. In the traditional Anglo–Saxon religions of the

# THE ENERGY PATTERN

British isles it was called the wyrd, an invisible network of magical forces. In theistic religions like Christianity, this oneness corresponds to what is called the Godhead, i.e., beyond the personal deity. In the systems language of post–modern science it is seen as an infinitely complex system of interrelationships, or 'web of life'. At the level of the planet Earth, this integrated whole is referred to as Gaia – the name of the ancient Greek Earth Goddess that has become the name of the whole Earth considered as a purposive intelligent living super-organism. (Id., p. 282)

Since we are part of the unified system of interdependence, just like every other being, we can never actually be outside of it, as a detached, objective observer. *But since the unified field is energy, we are energetically connected to every other form and being in the universe.* And since the field is also consciousness, this enables us, as human beings, to attune with, identify with, and communicate with any and every other life-form, object or being in the universe, from the macrocosmic to the microscopic. (Id., p. 283)

William A. Carey, a medical doctor who published amazing research on psychedelic substances in his book *In Search for Healing* (1996) for finding the missing link between *psyche* and *soma*, equally states that the human body, according to Edgar Cayce's prophetic visions, basically reveals to be an energy structure:

Cayce said, for instance, that this body we find ourselves in the moment is an *energy structure* and will respond positively or negatively to other energies; that the atoms and cells that make up the body are pure energy and have consciousness of their own. (Id., p. 4)

# The Language Pattern

In my own experience with *Ayahuasca*, as I reported it in my scholarly article *Consciousness and Shamanism (2015/2017)*, I made an amazing discovery about language:

> And the intelligence seemed to tell me telepathically that I was *locked in language*, that all my experiencing of life was *conditioned upon language*, and that I hardly ever perceived life directly, spontaneously, as an immediate connection.

I knew that this intelligence was connected to all, was connecting all and was all. It simply *was*, and it gently invited me to enter this connection, this all-encompassing love that it irradiated.

> The intelligence seemed to wanting to free me from the conditioning I had received through language, and through using language for describing reality.

What this extraordinary experience taught me was that language is indeed, as Freudian psychoanalysis strongly emphasizes, a *part of culture*, so that we can say that there cannot be culture without language.

However, what this intelligence conveyed to me was, so to say, the other side of the medal. Language is part of culture, but not forcibly part of nature's original setup. I might express it metaphorically the way that nature's language rather sounds like when, before the symphony starts, the musicians test and tune their instruments by playing the A unison.

Culture's language, it goes without saying, is the symphony itself, a written score, well defined, well put in print, well *black on white*—while the tuning of the instruments sounds like a form of chaotic improvisation and is volatile, and up to the moment, not fixated, not written down, not put in print.

Language serves social conditioning, or the other way around: conditioning is in our mainstream civilization done by language, by *verbal* language first of all, and by emphasizing the early acquisition of language skills by the young child. I certainly was conditioned that way, perhaps more than others; my mother told me that I already began to speak at the age of seven months and by the age of twelve months my verbal language capacities were as good as complete. My mother had been a radio speaker and her language training was very astute—which is a

# THE LANGUAGE PATTERN

great advantage for me today as a writer. But it also has strongly conditioned my perception of reality. I seem to perceive reality more than most others through language, and less through visual input, which is however the way most people function in our terribly visual culture.

I found that nature-abiding cultures have a much more general and encompassing notion of language. Language is understood primarily not as verbal language in the sense we use it in our culture but as a notion that encompasses the telepathic interchange with other language realms such as the animal and the plant realms. Language, in this sense, is communication in every possible way, be it by sound, be it by thought, be it by gestures or movements. It is clear that such a holistic notion of language as it seems to be common to most tribal cultures leads to a more subtle understanding of nature because it is *cross-lingual*, so to say. When I dispose only of human language, and spoke I twelve languages, I will still not be able to communicate with an animal, or with a plant.

The native may also want to master the languages of other natives tribes, but he will for that reason not neglect to master what I may call *the universal language* which is a form of sensitivity that includes telepathy and that is founded upon a deep interest in other realms of living.

Thus, I must ask, what does 'culture' in fact mean, if it is only a system of restrictions, or if it seems to be a *stupidity-factor* in human development, when I compare it to the much deeper wisdom of nature? Culture that defines itself as distinct from nature will use a distorted language. This kind of culture, which is ours, is created by perverted language, conditioned language and restricted language. In native cultures, which are *true cultures* in the sense that they integrate nature, I observe, the taboo is a restriction of action, but not a restriction of language.

When I see that, I understand that taboos that prohibit language are not only impeding communication, but destroy culture.

The prohibition of talk creates human beings that are mute, people who have to use other forms of communication than verbal expression. They then use the language of the fist: violence! This is the status quo of modern international consumer culture where worldwide terrorism had to come up in order to show even the most ignorant members of society what degree of collective violence we have attained. And we have attained it not by talking but by being and remaining mute! Tribal cultures, in their natural wisdom, know all this and care for the importance of language and the fact to *put words on things*.

Furthermore, in those cultures there are special festivities that are focused upon humanizing the taboo through its verbalization. Language humanizes the taboo and integrates it into the individual and the collective unconscious.

The spirituality of man is in first line its capability to humanize asocial behavior through verbalization and, doing this, to create culture. However, a taboo that represents a mere *non–dit* will not preserve culture, but destroy it, for the tabooed behavior cannot be avoided as long as it is not humanized by language. Any possibility to communicate it is cut–off by the talk-taboo.

Talk-taboos are the result of hypocrisy and undermine the taboo since what is not talked about is done secretly while it is denunciated publicly. Talk-taboos are thus against democracy since democracy implies free speech.

It is for this reason that language education and the support of the young to express their emotions and all what can be said is so overwhelmingly important for the formation and preservation of culture. This was historically the case in humanistic education which was based on the study of the old languages and the Hellenic tradition. These ancient cultures namely had less talk taboos than our today's mass cultures. After the end of the Hellenic era and the beginning of the *moralistic epoch of mankind*, namely under the influence of post-platonic and Christian thought, language was more and more tabooed and the possibility of complete dialogue more and more narrowed.

With good reason, psychotherapist Robert M. Stein calls Judeo–Christian tradition *primitive* in the right sense of the word, because it's undeveloped on a level of what I might call 'cultural cognition:'

### Robert M. Stein

> Creative psychological development, individuation, is dependent on spiritual freedom. When we say, for example, a man has a free spirit, do we mean that he freely or necessarily transgresses the imposed manners, mores and taboos of his culture? I think not. But it does mean the freedom to do anything or go any place he desires in the imaginal realm. He is a man who has

clearly distinguished the sacral, timeless world from the secular, historical world. He knows he can move with unashamed dignity among the gods and demons which belong to the mundane world. Such freedom cannot occur with a primitive form of consciousness in which inner and outer reality are governed by the same laws and values. In this sense, our Judeo–Christian tradition is primitive in that our thoughts and desires are subject to the same dogma, the same regulation, as our deeds. Spiritual freedom requires a break with biblical tradition and the development of a new form of consciousness – a consciousness which promotes the cultivation of imaginal freedom.

—Robert M. Stein in: *Redeeming the Inner Child in Marriage and Therapy (1990), 261 ff.*

Cultures that prohibit language, such as ours, are in truth no cultures as they are in their very root against freedom and against culture. They are Barbarian, authoritarian and tyrannical. Their well-sounding democratic setups and constitutions do not alter this fact and contribute rather to veil this truth.

The way to personal freedom and creativity, to autonomy and the detachment from collective categories is only possible through individually building a *culture of language*, the civilization of personal expression. Only on this very personal and individual basis culture can be created for the community, for language is every kind of expression, everything a creative mind may come up with, not only speech and writing, but also every form of art, of expression that leads to communication.

Our asocial instincts and our perversions can only be humanized through language; there is no other way. Without language the process that Freud called *sublimation of the instincts* is impossible. In case that a given society lacks language to express asocial behavior, *repression* takes place.

What is not integrated, is disintegrated, repressed and projected. All collective tragedies of mankind were and are accompanied by collective psychosis, a process that breaks through the fragile balance of repression and brings to the surface the primary archaic patterns of behavior. It is the same on the individual level. An ego without language is a psychotic ego and to create collective muteness means to prepare individual and collective psychosis.

The Tao of psychic health, then, not only for the individual, but also for entire cultures is the way of language, of communication and the active use of speech and writing. To realize this, it is not enough to

## THE LANGUAGE PATTERN

write well-sounding guarantees in national constitutions and otherwise shut up as a *politically correct* citizen.

An eminent expert on the importance of observing the formation of language in the human baby was the late Françoise Dolto (1908-1988), the famous child therapist from Paris, France, whom I have known personally. I visited her *Maison Verte* (literally means: *Green House*) in Paris in 1985 and interviewed Dolto after the visit in her apartment. An interesting correspondence followed up to our meeting.

Dr. Dolto was a pioneer of child psychoanalysis, made unique discoveries in this science and was known for her remarkable if not miraculous healing successes with psychotic children. In my observation, her success was due to a unique and almost *shamanic* understanding of language, which of course includes the direct telepathic language of our subconscious mind when it bonds in dialogue with another human's subconscious. In the daily practice of the *Maison Verte*, a center for parents and children, the difference it marks regarding conventional parent counseling becomes obvious when, for example, the child is greeted first when s/he arrives with their parents. *Children bring their parents!* Dolto used to say. The fact that the parents remain anonymous has been a very important detail in the good functioning of the place. They are identified as Jacques' mom or Helene's dad which was a very elegant way to protect the parents' anonymity while creating a language system that is child-focused. Without anonymity most parents would not have come to the group in order to discuss family problems.

Dolto's insistence upon language and truthful communication may sound strange in the ears of people who have never heard about psychoanalysis. To remedy this lack of knowledge, there is no better way than to read Dolto's extensive publications and to learn from her extraordinary penetration of the matter and her wealth of experience.

> —See, for example, Françoise Dolto, La Cause des Enfants (1985), La Cause des Adolescents (1988). See also my own study on her work, with book reviews and a good deal of accurate translations in: Peter Fritz Walter, Françoise Dolto and Child Psychoanalysis, Great Minds Series, Vol. 4 (2015/2017).

In the *Maison Verte* it is first of all the child that is spoken to, however not in the usual way people speak to children, but as one would speak to an adult. Babies are addressed as *Monsieur X* or *Madame Y* while they are shown around the facilities when they come for the first time.

Françoise Dolto revived an old tradition from the European aristocracy; in fact, more than two hundred years ago it was common in aristocratic families to address even small children with the same forms of politeness one addressed adults. Consequently children are not kissed or fondled or found 'cute' when greeted by the receptionists. Tenderness, such is the philosophy, has its place only in relations between parents and child.

In addition, the verbal communication with children is painstakingly truthful and serious, and no lies are told to children. When a child, for example, says it had no father, because the father left the mother, the center assistant would insist:

—But of course you have a father, although you do not know him. Both your parents have created you in their loving embrace, your mother and your father. Your father has participated in your procreation with his body. You see that you certainly have a father.

There are no truths of life that are hidden in front of children. By the same token, assistants are specially trained to avoid allusions that are unclear and subject to misunderstandings. Facts of life are named and explicated. Even though parents are discussing intimate things, children are not complimented out of the room if they want to stay with their parents. Nothing that concerns the child is discussed behind his or her back.

In many cases nothing but the respectful, rational, pragmatic and fearless way of handling relationships in the *Maison Verte* leads to solutions in behavioral problems with parents and/or children; fixed roles are thus easily dissolved and more flexible and creative forms of behavior are learned. The press often spoke of miraculous changes with people who attended the house.

One of the main goals of the *Maison Verte* was to prepare toddlers and small children for day care. Françoise Dolto, before she opened the center, had come to the insight that an immense number of children are traumatized by prematurely entering day care. Not only the child needs to be prepared, Dolto found, but also the mother. The sudden separation

# THE LANGUAGE PATTERN

from the mother for long hours of day care can represent a great shock for a small child.

Contrary to most day care institutions where the transition of the child into day care is facilitated by a gradual extension of the time span the child passes in care, the *Maison Verte* practices a different approach. Not the adaptation for separation is the decisive factor, but the *maturity of mother and child* for the separation.

Often, Dolto found, the children were ready for it, but the mothers not. The result was that the mothers gave to their children highly contradictory messages which were triggering emotional disturbances. For example a mother would affirm to the child that day care and consequently a separation was needed, but telepathically suggest to the child that she herself was not ready to stand the separation without immense emotional suffering. On the other hand children were found to be not mature yet for the separation when the mother had firmly decided to give them in daycare.

The *Maison Verte* insures that both, mother and child, are ready for the separation. This is of special importance within the current nuclear family structure where children are almost exclusively fixated upon their parents.

# THE LOVE PATTERN

## CULTURE AND PLEASURE

The present state of violence came about through wrong relationships, the sacrifice of love and the upsurge of morality and collectively regulated sexual behavior, and first of all by a deficient or totally lacking relationship of the inner parts of the psyche to each other.

> —See, for example, Wilhelm Reich, The Invasion of Compulsory Sex–Morality (1971) and Stone & Stone, Embracing Our Selves: The Voice Dialogue Manual (1989).

We do not just have violence. The problem is that violence multiplies because there are multiple causes that trigger violence, and they are adding up to each other. Succinctly speaking, if our problems were originating just from one single source, we would have since long solved them.

As a race, humans are presently learning that most effects have more than one cause; the idea of a straight line linking one particular cause to one particular effect is infantile. Yet mainstream science establishment is today still based upon this myopic view.

Only training our systemic and whole-brain thinking capacities can help us further.

## Pleasure–Denial and Violence

Violence is the result of a power vacuum that comes about through an inner fixation or complex within the *lower self* that acts as a compensation to suffering early in life. People who are in touch with their inner truth and who are liberated of culturally created fear blockages are able to realize constructive personal and collective happiness and they tend to build meaningful relationships.

Our love choices depend not on what our crime laws stipulate but what is ethically sound and viable.

More and more, it will be possible to make ethical and responsible love choices for relations that are unusual or even tabooed by former moral laws that belonged to the cultural heritage of the Pisces era. In that era of our collective past, happiness was smashed by multiple nonsensical ideological doctrines. It is these doctrines and their fierce dogmatism that have created the almost chaotic state of violence that we face today in the world.

One of the main objectives that flow from this insight is the urgent need to redefine *as natural* all erotic longings and sexuality for all ages through information and social reform. Part of this endeavor is to publish the well-hidden facts about the roots of violence and the current subtle manipulation of the credulous masses into a hyper-violence paradigm that will result in a more gigantic destruction that was ever seen before on the globe.

## Compulsory Sex Morality

Among the main reasons for violence being the repression of natural body pleasure and free love between people of all ages in general, and the child's free sexual life in particular, my task was to retrace the wrong turn that humanity has taken since prehistory and to embed this truth in a cross-cultural perspective that is focused upon the importance of love as a major factor of human evolution.

Compulsory morality or moralism has been the major killer app for love during all ages and it is moralism that brought about most of the current violence and destruction all over the globe. The pioneering work that Dr. Wilhelm Reich (1897-1957) has done in this field of research is of paramount importance. The essential truth gained from years of research on the functional processes of life is that all parts of the psyche must be given a voice so that a constructive inner dialogue can be set up.

Abuse is ill–defined in our culture. It only considers the victim and not the abuser. However, the abuser is a victim in as much as the person he has victimized. In fact, any other than non-violent and consenting love and sex interaction between two people, regardless of their age, simply is a lack of information and still more a lack of physical love experience. In addition, it is true that nobody can be victimized who has not previously chosen to act as a victim in a given situation. The abuser is trapped by the victim's paradigm in as much as the victim is trapped by the abuser's power problem. Both attract each other and there is no abuse without mutual implicit consent about acting out the two sides of abuse, the active and the passive one. Fighting against abuse is therefore not a moral cause but must start from a rational and two-sided view of the problem as an *entanglement situation* that is karmic and inherent in both parties' life matrixes.

Moral wars, by contrast, only lead to more confusion, more destruction and more abuse; for they do not tackle the roots of abuse that are the same roots as the roots of violence, but only are concerned with the reflections that such shortcomings produce on the surface of society. They are for that reason entirely ineffective and superficial.

A true remedy can only come from tedious study and observation of all the factors involved in abuse and those factors are for the most part unconscious elements of consciousness, entanglements that are hidden in the psyches of both abusers and abused, energetic blockages that have cut off the stream of life in one or the other way so that parasitic patterns came about.

There is thus an urgent need to change the reigning paradigm regarding love and abuse so as to reduce violence and to bring about positive change for constructive new relationships that are based upon the *golden rule of conduct* as it is taught by sages since times immemorial. It is to be seen in what horrendous way both clerical and politically fascist movements and leaders have since centuries tried to veil this essential truth and thus spread the *emotional plague* all over the globe.

After extensive research on mythology, particularly the writings of Joseph Campbell (1904-1987), I saw how the present love-killing paradigm has come about from ancient times and how it was possible that the former love-based world order was overthrown and violently eradicated by a world order that replaced love by morality and natural care by obligatory and institutionalized forms of family relations.

—See, for example, Joseph Campbell, The Hero With A Thousand Faces (1973), Occidental Mythology (1973), and The Power of Myth (1988), and See Riane Eisler, The Chalice and the Blade (1995), Sacred Pleasure (1996).

Historically, the transition in human prehistory from peaceful and life-affirming matriarchal fishing-farming cultures to violent and life-denying hunting-killing patriarchal cultures is of particular importance for the understanding of the present macho and hero paradigm with its strong moralism and all its life-denying and sex-repressing dimension.

There are important humanitarian consequences of this insight. Special care must be bestowed upon children who can be reformed and healed from biopathic deformations and character armors. There are millions of orphans in state institutions all over the world. If only a small percentage could be taken care of in collaborating with responsible institutions that understand the importance of the love paradigm, humanity would be served as a whole and true evolution would be possible! The spiritual or religious impact of this project is obvious; it would lead us back to our divine origins, through reconnecting to our higher self. This is true for all involved, the children, the educators and all those who help building this new worldwide educational system.

There were a great number of publications appearing in those years of the *hippie generation* that reported about experiments with freely raised children. What now seems highly disturbing in the *New Age of Fascism* that we are presently entering is the fact that those children were astonishingly mature and ready to assume appropriate responsibilities, highly flexible and adaptable to sudden changes in their life milieu, intelligent and independent.

> —See, for example Larry L. & Joan M. Constantine, Treasures of the Island (1976), and Where are the Kids? (1977).

Political restoration has spread its jovial and patriarchal wings over good old science, sweeping under the carpet what does not fit in the reigning restrictive and somewhat paranoid worldview. *Sexual permissiveness*, once a big word, was and is a myth in Western cultures!

> —See already J. P. Alston & F. Tucker, The Myth of Sexual Permissiveness (1973).

## Anthropological Evidence

In his book *The Invasion of Compulsory Sex Morality (1971)*, Wilhelm Reich referenced and discussed Malinowski's field studies on the Trobriand Islands in Northwest Melanesia, one of the few still existing matriarchal cultures.

> —See for a definition of what is understood under matriarchal culture the extensive studies of Johann Jakob

# THE LOVE PATTERN

> Bachofen, in: Gesammelte Werke, Band II, Das Mutterrecht (1948).

In fact, as early as in 1929, Malinowski published his report on the sexual life of the Trobriands in which he draws the reader's attention in particular to the undisturbed sexual life of children and adolescents.

> —Bronislaw Malinowski, The Sexual Life of Savages in North West Melanesia (1929).

Malinowski found, not without surprise, a *high sexual permissiveness toward children's free sexual play*. More generally, he noted the total absence of a morality that condemns sexuality in children.

> —Bronislaw Malinowski, Sex and Repression in Savage Society (1927), p. 76.

Instead, he observed that children engage in free sexual play from early age. Initiatory rites were absent with the Trobriands as children were initiated from about three years onwards, generally by older children, in all forms of sexual play. This play was non-violent and encompassed, with the older children, complete coitus. The most interesting finding for Malinowski was that in this culture *violence was as good as non-existing* and that there were as good as *no sexual dysfunctions*. Trobriands were found to be almost ideal marriage partners and divorce was a rare exception. Violent crimes were nonexistent and incest was strongly tabooed and inhibited by social norms and customs.

Other researchers found similar phenomena with the *Muria tribe* in South India where children stay until their maturity in so-called *ghotuls* where they live their sexuality freely and in utter promiscuity.

—V. Elwin, The Muria and their Ghotul (1947).

Older children initiate younger ones progressively into sexual play. These researchers found that after a phase of promiscuity, the children, from the age of sexual maturity, formed strong bonds and partnerships which were based not on sex, but on *love*. They further found that these first steady relationships formed the basis for later marriages that, regularly, lasted lifelong.

—V. Elwin, The Muria and their Ghotul (1947), Richard L. Currier, Juvenile Sexuality in Global Perspective (1981).

Some researchers and sociologists allege nowadays that these findings had no significance for our culture since they could not be extrapolated from their original cultural setting.

## THE LOVE PATTERN

However, such arguments assume that man, depending on his cultural conditioning, was basically different from one culture to the other. This is questionable, for the *biological foundations are with all human beings the same*, regardless of cultural or social conditioning. If all anthropological or psychological insights were valid only in a given culture, how could psychoanalysis which was founded by Sigmund Freud in Austria be successfully applied in Italy or France, in the United States or even in India or South America?

The truth is that those critics hide their own emotional blockages and blind spots behind pseudoscientific arguments.

One cannot disregard the extensive field studies of experienced anthropologists such as Malinowski or Margaret Mead or sweep them under the carpet with half-truths and moralistic philippics as it now is the trend, especially in the United States. Historically, the love pattern was integrated and lived by the majority of tribal cultures, but never was accepted by any of the larger dominator cultures that today form the core of our industrialized nations on the globe.

My thesis is that the destructiveness of civilization *is the result of the repression of the natural emotions of the child and the building of moralistic behavior structures* that have gradually replaced the primary self-regulatory processes that nature has coded for the growth of all living.

## Love Osmosis

Violence and destruction that characterize human history have their roots not in a biological or genetic error, nor otherwise in the human setup, but generally in the failure of man to keep in touch with nature's wisdom, and in particular by replacing love with compulsive morality and perverting children into obedient robots that have repressed their feelings in order to survive and be accepted.

Examples to the contrary, as already mentioned, are the pre-patriarchal high cultures of Antiquity, the Trobriand and other rather remote island cultures, and some more well-known cultures such as the Balinese culture, where people are generally emotionally balanced, happy and productive, loyal and intelligent.

# THE LOVE PATTERN

> —I speak here about the pre-Hellenic cultures such the Minoan Civilization on the island of Crete that was a highly developed civilization with a natural focus on the senses an on beauty, free sexuality and a matriarchal worldview, respecting the female and female children. It had a low crime rate, no slavery, no male god but venerating goddesses, and low degree of violence, a culture that however was raped and burnt down by the patriarchal invader tribes.

Crime rates in those cultures, if we take only the Balinese culture as an example, are extremely low. Violent crimes such as murder and rape, or the rape and killing of children, are as good as non-existing. Marriages are generally long-lasting and divorce rates are considerably lower than in modern society.

These cultures are more matriarchal in character than the highly violent modern civilizations which are predominantly patriarchal. History reveals that already the first highly developed civilizations, such as Sumer, Babylon or the Inca culture were early patriarchal systems. With patriarchy began the oppression of women and children and the reduction of sexuality toward certain 'acts' that were allowed and certain others that were prohibited. With the increase of power for the patriarchal system, repression, denunciation, intolerance and violence began to reign where before freedom, peace and tolerance were blooming.

An important factor within this process that keeps worsening until today is the repression of the child's emotional life.

## Love versus Morality

As already mentioned, *Bronislaw Malinowski*, a renowned anthropologist, found with his field research on the *Trobriand Islands* in Papua New-Guinea that this tribal culture has created social institutions that support the free development of the child's sexuality. This is through maintaining special houses for children and adolescents. From age three, children stay and sleep in these houses, together with other children, and live their child sexuality freely and in total promiscuity. The older children gradually initiate the younger ones into sexual relations, until coitus.

The Trobriands think that the child must live out his or her inborn sexual drive in promiscuity in order to be able, after puberty, to form steady and stable relationships with a partner for marriage.

# THE LOVE PATTERN

—Bronislaw Malinowski, The Sexual Life of Savages in North West Melanesia (1929), Sex and Repression in Savage Society (1927).

Malinowski was astonished to see that with the Trobriands marriages were indeed stable, the divorce rate being below five percent. Regarding violence and crime, it was virtually non-existing on Trobriand.

The way back to love can only start from the destruction of any and every concept of morality, be it justified religiously, ethically, culturally or scientifically. If we are to regain psychosomatic health and sanity, we have to stop thinking negatively about nature and accept our humanness without flinching, and also without the senseless zeal to improve what is, as it is. Without a morality concept, we again have to face the crude reality of *suchness* in everyday life. Instead of escaping in our comfortable realms of artificial duality, we have to practice *acceptance* when we realize that nothing is *per se* black or white, bad or good, unless thought declares it so. Where there is morality, love cannot be. Moralism and love are mutually exclusive as violence and pleasure are mutually exclusive.

Morality is associated with violence, love is associated with pleasure. Love and pleasure are the original ingredients of life, whereas morality and violence are decay products resulting from the sacrifice of love to the gods of respectability and materialism.

The price that civilized humanity pays for the destruction of love is high, as high as the price of life. Life on earth depends on our turning back to the wholeness, humility and permissiveness of love, abandoning the fragmentation, arrogance and rigidity of morality-based thinking and acting. As all pertinent research shows, love brings about natural growth, prosperity and happiness without effort and without strife. Most if not all our current worldwide problems could be solved peacefully and without conflict if we abandoned moralistic attitudes and hypocrisy in politics and based our strategies on love and cooperation. Instead of rigidly adhering to static concepts, we would understand the *unendingly dynamic and flexible* way by which nature acts and just like nature, we could bring about positive and constructive changes in every moment.

What I call *Love Pattern* is by no means an intellectual concept, nor is it an invention of mine. What I propose is looking at life with the eyes of life, instead of looking at it through the myopic glasses of morality. Anthropological evidence and cross-cultural research have clearly demonstrated that contrary to psychoanalytical doctrine, the oedipal confusion is not a universal psychosexual necessity in the development of human sexuality, but the inevitable result of love-denial in early childhood and the unnatural erotic fixation of children upon their parents.

What modern-day *child protection* does is advocating emotional incest, and thus what it gives to our children is entanglement, not freedom. Children who enjoy their sexuality by having sex play with other children and adult mates are not confused about their emotional attractions and sexual preferences. Much to the contrary, these children have a mindset that, compared with their virgin comrades, is based upon what Freud called *reality principle*. Under the definition of modern child psychology, they would have to be considered more adult than most of our current adults that are raised like infantile idiots. They are in fact more responsible, more considerate, more socially minded and less selfish than the prototype of the castrated consumer child.

In my discussions of this topic, I regularly get the question why, if this was so, society was so hard on admitting and accepting the existence of *pedophilia* as a natural counterpart to children's own love wishes? My answer is that this society is so deeply ingrained in incest in every possible form that admitting children's emotional and sexual freedom would jeopardize the expectation of most parents to be the foremost and exclusive love mates of their children. There are many people in our culture who think that allowing the child *to be sexual* would naturally encompass children having sex with their own parents. When I object that children, if let free to choose their sex partners would in most cases not opt for their parents but for peers or adults other than their parents, people seem to be puzzled in a really interesting way. I namely see in those moments that they are *more* puzzled and disturbed by the idea that children have sexual relations with strangers than with their own parents.

They are in fact more sympathetic to the idea that children have incestuous relations rather than free choice relations outside of the family web! And this is, sorry, really a perverse idea and it shows the deeply perverse base setup of our patriarchal society that considers children first of all as pleasure toys for their parents and extended family, and only *afterwards* as people like you and me who have the right to choose their partners instead of complying with the unwritten law to serve as kiss cuties for consoling sexually frustrated elders that happen to be their parents.

After all, let us ask the pertinent question: Why should parenthood grant sexual privileges? My guess is that when one day we have enough discussed, publicized and mourned about rampant incest within the modern nuclear family, and all the cards are on the table, the taboo on pedophilia will be lifted, because it will be seen that it's after all more natural to have children copulating with choice partners than to tie them as sex dummies to their parents.

It is noteworthy to observe that people who reject children's right to be sexual and to live choice relations rather than being night pillows for their neurotic parents *are often sexual virgins* and as such sexually as little experienced as most consumer children. In addition, by actively defending incest, such individuals may unconsciously strengthen the repressive *child-abuse paradigm* of the majority and defend parental interests in controlling and manipulating children, instead of serving the true interests of children for *autonomy and self-regulation* in love and sexual matters.

Abandoning morality and returning to love will not result in upholding incest, except in the rare cases that the sexual interaction between parents and children is mutually agreed upon and shared as a conscious bonus in the parent-child relationship. In the regular case, incest serves but the parental interest for affirming and re-affirming control and dominance, and leaves the child little space for yes-or-no decisions.

Whatever one may think regarding this subject, nature has given us millions of potential love mates, and the moment we choose sex *within the family* instead of *sex within the world*, we show that family life has an undue dominance over us and that we have not made the step from the cradle into the world —which is a world of free choice and not a world of freedom within a lion's cage.

Interestingly, outside of the Western world, I never met a child who upon my question affirmed of wanting to engage in, or having engaged in any sexual affair with a parent or close relative. To most of these children, the very idea of incestuous sex is clearly negative or not even occurred in their mind because they were busy and satisfied with love and sex relations outside of the family.

With prostitute children in Asia I found that these children clearly distinguish between the relationship they maintain with their father, on one hand, and that they wish to be platonic, and the erotic relationships they maintain with their male sex clients, on the other hand.

Incest is nothing bad in my view, nor do I believe that it's per se immoral. And yet it is not a viable option when millions of other love partners are available. The deeply unhealthy codependent attachment that *emotional incest* typically results in is clearly counter-productive to building autonomy and personal strength in our youth. Instead of advocating more incest, we should advocate more erotic love options for our children *outside of the family*.

## Rebuilding Trust

Abandoning morality and returning to love means that we begin to trust the dynamic self-regulatory processes that are built into all our life functions, and our natural biological relationships. As most Westerners have desperately lost their continuum and with it their innate trust in the goodness of nature and natural living, this is the crucial point for most of them.

As I was just like most of them some two decades ago, I know what I am talking about. But I also know that this trust can be regained, even if we were suffering from a deeply negative and humiliating childhood and youth. Without trusting in nature, we do not know what it means to trust another human, and without this basic trust, love between humans is impossible. It is as simple as that. All our senseless rhetoric about *child abuse* shows that we do not trust nature and therefore mistrust the natural self-regulatory wisdom of love.

Despite abundant research showing us that one of the results of this mistrust is violence, we continue to put our trust in authority and the nonsense of tradition instead of trusting what is in front of our eyes. All our sages told us that draconian laws only show that the natural law of love has been lost, but we continue to put every year more people in jail than we

## THE LOVE PATTERN

liberate from our prisons thus discarding out a steadily growing part of our societies.

How can we continue to declare more and more people socially inept while we as a society engage worldwide in wars and massacres that defy any of the atrocities that we blame our prisoners with?

Building trust is the first step if we are serious to regain love as our foremost regulatory life principle. Here, our scientists and our poets for one time agree that this move is a good one and that nothing can defeat us if we put our trust in love.

# The Pleasure Pattern

An intelligent society tolerates to a certain extent, and socially codes, perverse behavior in the sense that it considers *violence as the only true perversion*, while it can comprehend that unnatural sex ultimately is human. That is why most of the ancient wistful civilizations that were not perverted by the plague of moralism had social outlets for pedophile, incestuous and other marginal sexual longings.

The human nature only knows *one* taboo, that is, doing harm to another. And violence is what most of the time causes harm, not perverse sex. This is simply so if present–day moralists and world puritans accept it or not. Culture is not always in accordance with nature; and the human being has the unique ability to forge culture that obeys to its own *cultural laws* instead of abiding by the laws of nature. I do not say that this should be so, and I do not judge what ought to be or not be in terms of human behavior. What I do is to observe and describe what I see; and what I see is that culture can well be different from nature's wisdom. Our patriarchal culture is against nature and yet it has survived until this day. But it will ultimately probably not survive because it's not intelligent to build a culture that turns natural laws upside down. Yet I think that our admittedly flawed patriarchal culture, had it not a definite hangup with fundamentalism, and had it integrated female wisdom and power and tolerated sexual perversion as the *little crime* in its various forms as unnatural and incestuous sex that serves to defend the *big crime*, that is structural and domestic violence, could survive because of its ability to flexibly adapt to new situations and environments.

Françoise Dolto was noted to say that the fundamental difference between perversion and neurosis is that the former one day is dropped by satiation, as the energy in perverse behavior is to be naturally exhausted one day, so that the perverse habit is being dropped, whereas neurosis only worsens over time if no therapy is engaged.

This is so because in perverse behavior there is still a good portion of nature's flexibility, which is why the behavior is not rigid and can change. The same is not true for neurotic behavior. The ultimate issue, then, with patriarchy is not that it's perverse and against nature, but that it's neurotic! In fact, there is a natural logic in every pleasure-seeking behavior.

For example, men who seek complete love relations with little girls or boys, while mainstream society considers their sexual longings as perverse, can one or the other day outgrow their immature sexual attractions and *grow up* sexually, and as a result turn to older partners. Or they may not, considering their admittedly childish sex as a form of poetic lifestyle. And it is. You cannot copulate with a 2-year old the same way you can copulate with a 20-year old. Your sexual behavior will thus be naturally restricted by the smaller size of the child's genitals, and if you want to bear up with this restriction, you are either a fool or you accept this limitation for the charm that you derive from mating, in a very incomplete fashion, with sexually immature people.

This is the reason why pedophilia, while its etiology may be known and understood, is generally not considered an *abomination* by conscious pedophiles themselves, while society well considers it that way. This is because the pedophile will not define his or her behavior with its sexual part only, or based on their sexual longings, but within the greater context of their *emotional attraction* to children, and the poetic world that they build around their relationships with children.

And for good reason! Our society has namely destroyed much of this poetic world that not only pedophiles love and cherish, but generally sensitive men and women, and particularly our gifted poets, writers, musicians and geniuses who most love in life what is genuine, innocent, young and exuberant!

Sexually engaging with a person who is under a certain age defined by the law is not a divergence from the behavior code that is so fundamental that it should be demonized socially and considered criminal behavior.

Succinctly speaking, societies, like present-day international consumer culture that develop into a Barbarian horde of judgmental persecutory terminators will pay a high price for their lack of *erotic intelligence*; they will destroy themselves in the long run by breeding such a high level of structural and domestic violence that they will one day suffocate in it. There is only *one* perverted form of pleasure: it's violence. Violence is the only true perversion in the human setup, while all sexual longings, however they may be considered by a particular cultural setup, are still connected with the healthy base layer of living, because they are not violent per se. These desires only *become* violent through their individual and societal repression because repression distorts and perverts the strong *élan vital* contained in these desires that, when unrepressed, is *partially sublimated* into art and poetry.

However, as I said already earlier, sublimation only works when consciousness allows the acceptance of the desire, and not when consciousness represses the longing by judging the particular behavior bad and socially destructive. Pedophilia, to stay with my example, is not socially destructive but its very repression is. While pedophile desires in their sexualized form may not be the result of a healthy upbringing, and while they may be considered forms of pathological sexual wishes, this does not mean that they have to be declares as 'criminal,' and that they have to be repressed. As long as they do not turn into violent rape desires and are not acted out as child abduction, rape and murder, they should be socially tolerated and coded as acceptable, while admittedly marginal forms of erotic social conduct. But our society does not follow the precepts of erotic intelligence, that's the problem here. I would say it creeps along the lines of emotional stupidity, and this especially since the last three hundred years of *enlightenment*, during which it has mechanized and robotized pretty much everything that was filled with soul and a certain poetic content—such as, for example, relationships with children. To be true, today these relations, in mainstream culture, are neurotic, standardized and idiotic, hypocrite through and through. And they are dominated by a denial of the body, by a denial of touch, and by a denial of truth, and of truthful language. This denial is based upon amnesia, the amnesia namely of our own childhood, with its insecurities, its dangers, and its secret erotic adventures. Most of us, in a deeply neurotic society, have forgotten that our bodies were the first and certainly the most natural source of pleasure. As a result, most adults in industrialized societies live lives that are not their own. Alienated from our bodies, we compensate for the lost paradise of *Being* through *Having*, possessing, consuming, to paraphrase Erich Fromm.

# THE PLEASURE PATTERN

—See Erich Fromm, To Have or To Be (1976).

The things we are attached to give us a cheap *ersatz* of the joy we could have if we had kept true to what is freely given by nature.

This dilemma begins in early childhood. The luxury of civilization costs a high price. We pay for it with our bodies that we gradually destroy. A body that is not connected to a soul is a dead body. If the soul is still in that body, death occurs while we live, in the form of alcoholism, drug addiction, heart disease, rheumatism, leukemia, cancer, immune deficiency disorder or other so-called lifestyle diseases.

The process of alienation that leads to this gradual decay of the human body is an integral part of the conditioning for the consumer society. Consumption that makes the prosperity of this form of human social life functions only if people actually consume.

This is not a communist view or opinion, but a socio-economic and verifiable fact. Modern society is organized that way, if we personally agree or not.

One of the most important conditioning devices are children's toys. The toy industry sustains the entire structure of the consumer society. Without the early conditioning toward toys as a *body pleasure ersatz* people would not accept the later ersatz satisfactions that they receive for the sacrifice of primary body pleasure.

What is primary body pleasure? It is the pleasure that already the small child derives from playing with the body. This pleasure is an essentially *sexual pleasure.*

The moral and societal prohibition of child sexuality is the primary condition for the functioning of a civilization of *ersatz satisfactions*. This fact explains the research results that prove a direct correlation between the civilization standard of a given society and the severity of its child sex taboo.

However, most scientists who have arrived at these conclusions have failed to see the *impact of this insight* upon human civilization and culture.

Freud thought that man develops culture through sublimating primary body pleasure, and thus not through the satisfaction of our innate desires; he assumed that culture was the outcome of a transformation of original libido energy into a form of creative energy that serves cultural purposes.

## THE PLEASURE PATTERN

Is this thesis true? I think it is true, but not in the sense Freud thought it was. It is true insofar as the prohibition and transformation of instincts leads in fact to a form of culture, an *ersatz culture* for the original culture that would have been created through living our original instincts. But it is not true in the sense that sublimation led to a true and original culture. After deeper analysis, we can observe that the denial of living our original desires leads to a denial of living our original creativity. The result is exactly the *fake culture* that today our international consumer culture represents.

By contrast, if we look, for example, at *Minoan Civilization* that did not repress sexual pleasure, neither in children nor in adults, we see how high human civilization can grow on the basis of an originally permissive attitude toward our primary instincts.

In many ways Cretan culture, which was not yet a patriarchal, but an *egalitarian culture*, was superior to our modern culture, more developed, more refined and, first of all, more peaceful and more sensitive toward our real human needs.

—Riane Eisler suggests to abandon the Bachofen dichotomy of 'matriarchal–patriarchal' replacing it by 'egalitarian–dominator,' thus avoiding endless discussions if or not in matriarchal cultures males were oppressed by females. The question in fact is not who dominates whom, but if a culture in general runs on a dominator paradigm or on an egalitarian paradigm. It is now shared by the majority of scientists that what we formerly called matriarchal cultures were clearly more egalitarian than the subsequent patriarchal or dominator culture. Thus, a way back to love obviously will have to consider a sort of *archaic revival,* to use the expression of Terence McKenna.

The sudden and brutal end of this immensely creative culture through the invasion of the patriarchal Phoenicians was one of the turning points in human history. It was this turn into patriarchy and violence that put humanity on the track of pseudo-culture; it was from this time that the artificial and hypocrite, the stupid and doctrinaire, the false and arrogant, together with violence, war and destruction began to dominate the natural and naturally intelligent original cultures that preceded them. All our great religions have absolved and fueled this turn into a *manipulative and undemocratic pseudo-culture* that still represents present-day mainstream culture. Religions played the role of a catalyzer in our conditioning for war and destruction—although they generally preached the contrary.

## THE PLEASURE PATTERN

It is significant that tribal cultures who put the human body and body sensitivity in the foreground of cultural, artistic and social life and where love is not faked, do not need to preach love, do not need to discuss love and do not need to heal love—simply because they *practice* love. Their religion is not the integrity of pseudo-moralistic values, but the integrity of *love*. Religion, in these cultures, is not a power factor and does not feed on power nor does it exert power over individuals. Religions, in these cultures, do what they should do in accordance with the true sense of the word *religio*, that is they give guidance to those who are searching for the truth about coming and going, transcendence of suffering, care for the sick, the needy, the marginal and the dying. The North American Indians, for example, have preserved forms of this original and most pure religion that was once universal for all human beings.

Herbert James Campbell, an English neurologist, found in twenty-five years of research a universal principle which dominates our brain: *the pleasure principle*. This sounds like Freud, but it has little to do with psychoanalysis or psychology. What we are facing here are facts proven by natural science, by neurology. Campbell' book *The Pleasure Areas (1973)* which represents a brilliant summery of years of neurological research. Campbell succeeded in demonstrating that our entire thinking and living is primarily motivated by pleasure, not only as tactile or, sexual or sensual pleasure, but also as intellectual or spiritual pleasure. With these findings, the old theoretical controversy if man was primarily a biological or a spiritual being, became obsolete. For it is in the first place our striving for pleasure that induces certain interests in us, that drives us to certain actions and that lets us choose certain professional pathways.

During childhood and depending on the outside stimuli we are exposed to, certain *preferred pathways* are traced in our brain, which means that specific neural connections are established that serve the information flow. The number of those connections is namely an indicator for intelligence. The more of those preferred pathways exist in the brain of a person, the more lively appears that person, the more interested she will be in different things, and the quicker she will achieve integrating new knowledge into existing memory.

## THE PLEASURE PATTERN

High memorization, Campbell found, is namely depending on how easily new information can be *added-on* to existing pathways of information. Logically, the more of those pathways exist, the better! Many preferred pathways make for high flexibility and the capacity to adapt easily to new circumstances.

Campbell's research indicates that the repression of pleasure that is part of our Judeo-Christian culture, has negatively infringed upon human evolution and impaired the integrity of our psychosomatic health. This is exactly what Wilhelm Reich found—without having had at his disposition Campbell's later neurological findings.

Not only neurologists such as Campbell have thought about the basic functions of life and living, but also people who were formerly active in totally different fields of science. The American scientists Ashley Montagu and James W. Prescott came from different research positions and perspectives. Montagu wanted to know why in animal experiments small rhesus apes died when they were deprived from their mother while they survived when a simple felt mat was put in the cage as surrogate of motherly tactile affection. Prescott researched on the origins of violence. He did from the start not acknowledge the age-old presumption that man was *per se* a violent creature even though human history, or what historians saw of it, seemed to prove it.

Both scientists basically came to the same result, that is, *tactile stimulation of the infant* is a main source of early pleasure gratification and a condition for human health, for harmony, and for peace. Ashley Montagu's research developed a specific focus on the importance of the human skin as a primary pleasure provider.

*Grant's Method of Anatomy* defines the skin as the most extended and the most varied of all our sensory organs. Ashley Montagu's study *Touching: The Human Significance of the Skin (1971)* was the final result of thirty years of skin research, not only Montagu's, but of others whose research Montagu evaluates.

Ashley Montagu's research is interesting as to tactile stimulation in early childhood. Montagu's specific focus during his research was upon the mammal mothers' licking the young. He found astonishing unity in zoologists' opinions as to the importance of motherly licking for the survival of the offspring. He discovered that it was in the first place the perineal zone of the young that the mother preferably and repeatedly licked. Experiments in which mammal mothers were impeded from licking this bodily zone of the young resulted in functional disturbances or even chronic sickness of the genitourinary tract of the young animals.

Montagu concluded from this research that the licking did not serve hygienic purposes only, but was intended *to provide a tactile stimulation for the organs that were underlying the part of the skin that was licked.* (Id., pp. 15 ff.) However, he further concluded, licking hardly ever occurs in the mother-child relationship with primates or humans. (Id., p. 18). With one exception: an Eskimo tribe, the *Ingalik*, was found where the mother licks the face and hands of the baby in order to clean them, until the baby is old enough to sit on the bench. (Id., p. 234)

Most researchers found that with progressing evolution, licking was gradually replaced by eye or skin contact between mother and child. The tactile needs of the small child seem to correspond to the desire of the parents to express love through *tactile affection* such as kissing or fondling, or pressing the child's naked body against one's own during sleep or rest, which is very common with Eskimos or Indian tribes.

In the run of industrial civilization, however, this has changed fundamentally. Modern pediatrics or child psychologists recommend parents to put their children in separate rooms and beds so that parents and children are physically separated. This is the main reason why the *civilized child* gets much less of tactile stimulation in early childhood than children from tribal cultures. A direct relationship was discovered between early tactile stimulation and the immune system of the child.

This relationship was corroborated by France's first and foremost obstetricians, Frederick Leboyer and Michel Odent.

—See for example Frederick Leboyer, Birth Without Violence (1975).

Michel Odent writes in his book *La Santé Primale* (1986):

> It is not yet completely understood that sensorial perceptions at the beginning of life can be a way to stimulate the 'primary brain,' at a time when the 'system of primary adaptation' is not yet grown to maturity. More

specifically, this signifies for example that, if one fondles a human baby or an animal baby, one also stimulates his immune system.

—Michel Odent, La Santé Primale (1986), p. 24 (Translation mine).

Montagu states that love was once defined as the harmony of two souls and the contact of two epidermises. In this sense the *peau à peau* that is nowadays is again recommended by pediatricians, is a *primal condition* for the healthy growth of children, the good functioning of their immune system and the early creation of *preferred pathways* in their brains.

Tactile stimulation of the skin thus can be said to promote high intelligence! This is really something our society has never until now understood, and that even most scientists do not acknowledge, despite the abundance of research that clearly corroborates this truth. If this was really understood, our need for tactile pleasure, shared nakedness and sex, however sex is acted out and however it is like, could never be subject of criminal law!

In his research with rhesus, Montagu reached astonishing findings. When he deprived the newborns of their mother and let them in the *naked* cage, they died. When he did the same, but put a kind of felt mat in the cage, they survived, although they carried away some brain damage from the deprivation of the mother. However, it was a fact that the *felt mat* assured their survival.

How could that be? Montagu went one step further. He replaced the mother through a *felt mother* that was hung in the cage. Now the young did not only survive but they also had almost no more brain damage. It was especially that the young survived simply by a felt mat being put in the cage that intrigued Montagu. Further observations led him to realize that the young rhesus used the *carpet to give to their bodies tactile stimulation*, which obviously served as a compensate for the tactile stimulation they normally got from their mother in form of licking. The interesting fact about this experiment is that it was not the milk of the mother nor her care that was essential for the young's survival, but exclusively some form of *tactile pleasure*. The felt of the carpet was similar to the mother's fur and therefore acceptable for the young as a mother surrogate.

This research amply demonstrates how important tactile stimulation is with mammals, and so much the more with humans where primary symbiosis is even more prolonged. James W. Prescott's research particularly focused on the consequences of early tactile deprivation. In his article *Body Pleasure and the Origins of Violence (1975)* Prescott uses R.B. Textor's supra-cultural statistics in order to scientifically prove his collective and political conclusions. Already in the 1930s, Wilhelm Reich disproved the very widespread misconception that sadistic and destructive tendencies were part of human nature. Reich strongly opposed Freud and his theory of a death instinct, stating that those destructive instincts were but secondary drives, *a direct consequence of the cultural repression of the natural sexual instinct* which had brought about a collective neurosis in the human animal. Reich's findings, at the time violently opposed by the majority of his scientific colleagues, are now confirmed by Prescott's research which brings statistic evidence as to the malleability of the human individual through his early tactile experiences or the absence of such experiences:

> Recent research supports the point of view that the deprivation of physical pleasure is a major ingredient in the expression of physical violence. The common

association of sex with violence provides a clue to understanding physical violence in terms of deprivation of physical pleasure. (…) Although physical pleasure and physical violence seem worlds apart, there seems to be a subtle and intimate connection between the two. Until the relationship between pleasure and violence is understood, violence will continue to escalate. (Id., pp. 10-11).

It is interesting what Prescott wrote in the introduction to his study:

Unless the causes of violence are isolated and treated, we will continue to live in a world of fear and apprehension. Unfortunately, violence is often offered as a solution to violence. Many law enforcement officials advocate 'get tough' policies as the best method to reduce crime. Imprisoning people, our usual way of dealing with crime, will not solve the problem, because the causes of violence lie in our basic values and the way in which we bring up our children and youth. Physical punishment, violent films and TV programs teach our children that physical violence is normal. (Id., p. 10)

Prescott thus fully confirms Reich's research and corroborates his socio-economic and sex-economic findings. More specifically, James W. Prescott found a noteworthy relationship between pleasure and violence. Referring to laboratory experiments with animals, he could detect a sort of *reciprocal relationship between pleasure and violence*, that is

# THE PLEASURE PATTERN

the presence of pleasure inhibits violence – and *vice versa*. Prescott states:

> A raging, violent animal will abruptly calm down when electrodes stimulate the pleasure centers of its brain. Likewise, stimulating the violence centers in the brain can terminate the animal's sensual pleasure and peaceful behavior. When the brain's pleasure circuits are 'on,' the violence circuits are 'off,' and vice versa. Among human beings, a pleasure–prone personality rarely displays violence or aggressive behaviors, and a violent personality has little ability to tolerate, experience, or enjoy sensuously pleasing activities. As either violence or pleasure goes up, the other goes down. (Id.)

Further, Prescott found a direct relationship between the *childrearing paradigm* of a given culture, and the degree of violence that reigns in that culture. In detail, he found that societies that tend to rear children in a rather Spartan way, hostile to pleasure and with little or no tactile stimulation, cherish in their value system various forms of violence, do warfare, torture their enemies, practice slavery and progeny and concede to women and children a rather low social status; these societies also exhibit a high crime rate. (Id., p. 12)

Another violence-indicating parameter in a society, Prescott found, is *physical violence towards children in form of corporal punishment. (Id.)* Further, repression or tolerance of children's sexual life plays a decisive role in the assessment if a society has a high or low violence potential:

> Thus, we seem to have a firmly based principle: Physically affectionate human societies are highly unlikely to be physically violent. Accordingly, when physical affection and pleasure during adolescence as well as infancy are related to measures of violence, we find direct evidence of a significant relationship between the punishment of premarital sex behaviors and various measures of crime and violence. (Id., p. 13)

As a result of his research, Prescott advocates the abolishment of corporal punishment of children, a genuine rise of the social status of women, the reinstitution of the extended family, the reintegration of the elder and a more active participation of men with *childrearing and the bestowal of physical affection on children* in their role as fathers, uncles or educators.

> —See James W. Prescott, Deprivation of Physical Affection as a Primary Process in the Development of Physical Violence (1979), pp. 77-78.

I discovered the writings of James W. Prescott, PhD in the 1980s, at a time when I was doing research on Ashley Montagu, Frederick Leboyer, Michel Odent,

## THE PLEASURE PATTERN

Alexander Lowen, Bronislaw Malinowski, and Margaret Mead. The two major articles written by James W. Prescott were coming to me like a revelation to a question I had asked since more than a decade: 'What are the roots of violence?'

Knowing from anthropological, ethnological, and sociological studies as well as from neuropsychology and from spiritual work that violence is not the natural condition for humanity, but a sort of emotional and cultural perversion that results from deep hurt suffered early in childhood, and probably also from scars that go back to former lives, I was grateful to have found at last conclusive research evidence that not only analyzed our condition, but also pointed to viable long-term solutions for creating a more peaceful society in the future.

Prescott's research also integrates findings by lesser known researchers as Dr. Harlow who have focused on the brain development of rhesus, and who found revealing evidence for the fact that among all the factors that make a mammal infant survive without the mother, the *one single essential factor is the availability of a 'touchable' object that provides tactile stimulation.*

For example, in a widely documented experiment, two mother surrogates were hung in the cage, one serving as a milk provider, the other being a soft doll made from linen. The surprising thing was that all rhesus infants preferred the *cloth mother* over the milk-giving mother, thereby signaling that tactile stimulation was the most important in their parenting needs, not the secondary availability of mother milk.

Today, this research has been corroborated by newer brain research, conducted by a variety of researchers starting with Herbert James Campbell in the 1970s, and with James W. Prescott as the expert who shows in a number of publications that tactile stimulation of infants together with breastfeeding and baby-carrying are the most important factors for building nonviolent, socially positive and non-abusive behaviors.

To repeat it, the solutions that James W. Prescott suggests are long-overdue changes in the process of childbirth and our educational system, permissive and nonviolent childrearing together with greater social permissiveness for premarital sex and a definite legal prohibition of physical punishment of children in both the home and school together with effective government collaboration for fighting domestic, educational and sexual violence.

Regarding infant care, Prescott stresses the importance of the *primary symbiosis* between mother and infant during the first 18 months of the infant, abundant tactile stimulation of infants and babies, using techniques of child massage, as well as co-sleeping between parents and small children.

Another important field of research that could be classified under the header of 'ritual violence' is both male circumcision and the widespread genital mutilation of female infants, girls and women, which is now discussed under the header of 'female genital cutting' or FGC.

James W. Prescott advocates the complete abandonment of such practices that I heard about first in 1984, when doing a legal research on the matter for Edmond Kaiser, founder of *Terre des Hommes* in Lausanne, Switzerland. At the time I thought these violent practices were limited to some communities in Somalia, Sudan and other African countries, but fact is that it's a worldwide problem. The *American Academy of Pediatrics* writes in their policy statement that it was estimated that 'at least

100 million women have undergone FGC and that between 4 and 5 million procedures are performed annually on female infants and children, with the most severe types performed in Somalian and Sudanese populations.'

> —American Academy of Pediatrics, Policy Statement—Ritual Genital Cutting of Female Minors: http://pediatrics.aappublications.org/cgi/content/full/125/5/1088.

In addition, what is lesser known is the fact, reported by the American Academy of Pediatrics that these practices are not limited to Muslim populations but are known also from orthodox circles among Christians and Jews.

The perhaps most important research topic where James W. Prescott is widely recognized as an expert is violence prevention. He particularly stresses the importance of breastfeeding and bonding for 2.5 years or longer. He emphasizes that nonviolent behaviors develop as a result of *cognitive affectional bonding* between mother and infant. Together with a number of other researchers, he has recently documented and published scientific evidence that shows beyond doubt that the human brain develops differently in humans who as infants have enjoyed prolonged breastfeeding, and in those who have not. It is interesting to note that the suggestions that James W. Prescott comes up with from his perspective as a peace researcher are very much in accordance with those suggested by Jean Liedloff, in her book *The Continuum Concept (1986)*, from her perspective of the lifestyle of native peoples.

Also, there is a striking similarity of solutions offered for the same questions by Ashley Montagu, as a result of skin research, and by the French obstetricians Michel Odent and Frederick Leboyer who have looked beyond the fence of obstetrics and into what Odent called *Primal Health*, which is a holistic concept of health and well-being.

In my perspective and the overview I had over Prescott's research, it seems to me that the central focus is the preparation of far-reaching policy changes for the political agenda that are backed up by hard scientific data. In so far, I consider Prescott as a researcher more important than many others who are perhaps more published and more famous than him. In fact, the importance of his research can hardly be underestimated. We are living wrongly as a society and the violence we face is not hazard, nor a 'biological mistake' but the precise result of our living against the wisdom of nature.

Research in neuroscience delivers the clear-cut evidence that touch is paramount for the development of nonviolent and socially positive behaviors. Dr. Prescott shows that sensory deprivation results in behavioral abnormalities such as depression, impulse dyscontrol, violence, substance abuse, and in impaired immunological functioning in mother deprived infants. He demonstrated through a research with 49 native cultures that there are precise correlations between *low affectionate cultures, insufficient mother-infant bonding, patrilinearity, polygeny, warfare, slavery, torture of enemies, sexual repression, child abuse, violence and monotheism*, on the one hand, and *high affectionate cultures, nurturant mother-infant bonding, matrilinearity, low polygeny rate, absence of warfare, no slavery and no torture, sexual permissiveness, high infant indulgence, peaceful coexistence and polytheism*.

To summarize, Prescott's research sees the primary problem in the etiology of violence in *failed bonding* in the mother-infant relationship and so-called somatosensory affectional deprivation (S–SAD), as

## THE PLEASURE PATTERN

such deprivation causes developmental brain abnormalities. The brain that results from this abnormal upbringing is the *NeuroDissociative Brain*.

It is related to *pain, theistic religions, gender inequality, sexual puritanism, addictive synthetic drugs, authoritarian control, depression, violence, warfare, a biomedical health model, and politics of betrayal*. The healthy brain, which develops when affectional cognitive bonding between mother and infant was nurturant and adequate, is able to experience pleasure. It is related to *earth religions, is matrilineal and favors gender equality, sexual liberty, natural botanical drugs, egalitarian freedom, a biobehavioral health model and politics of trust*.

It is important to realize that we have not one single factor here, but a whole *pattern of factors* that belong as it were together. This is exactly what I emphasize in my own research on the *Eight Dynamic Patterns of Living* in that I show that most native cultures that are allowing to build the limbic-subcortical emotional brain through adequate parenting are favoring eight patterns of living in their overall lifestyle, which are autonomy, ecstasy, energy, language, love, pleasure, self–regulation and touch.

# The Self-Regulation Pattern

Oedipus, in Sophocles' famous tragedy, killed his father and married his mother. He didn't do this consciously but was led by the invisible threads of fate or destiny. The whole tragedy was triggered by an oracle that told Oedipus' father he would die through the hand of his own son. Wanting to escape the impending fate, the father, after his wife gave birth to a baby boy, let bind the infant, pierce his feet, and told a slave to abandon him on top of a mountain. The slave, however, did not follow the order and, being moved by the suffering of the baby, took the child home from where, through a chain of coincidences, the baby eventually was being taken to the King of Corinth and his wife, who formally adopted the child and called him *Oedipus*, a name that means *swollen foot*.

When Oedipus was grown up, he left in order to find out about his destiny, as some people had made remarks that he did not for the least resemble his parents. Without however knowing the true story about his origins, as his adoptive parents never had revealed it to him, Oedipus traveled to Delphi to ask the oracle and god Apollo warned him he would kill his father and marry his mother.

Oedipus thought he could easily escape the prediction simply by not returning to Corinth, to his supposed parents. Yet his escape was illusory; it was through Oedipus' solving the Sphinx's riddle that he was going to marry his own mother. The marriage as well as the patricide that preceded it were acts that Oedipus committed innocently, not knowing the truth. He was thus not aware of what he was doing. Or, to put it in other words, he was completely blind regarding the truth of his fate, of his life. However, as the story continues, it becomes clear that, despite his innocence, Oedipus was *not* considered free of guilt.

This strange story became the basis of Freud's theory of the *Oedipus Complex* which is an angular point of his theory of infantile sexuality. Furthermore, anthropological studies have shown that the Oedipal problem is universal in all cultures that repress the free sexuality of the child, but *only in those*. Needless

# THE SELF-REGULATION PATTERN

to mention that on Trobriand there is no Oedipal problem or complex to be found in the psychosexual development of children. Equally universal are the taboos of incest and patricide/matricide. However, cultures differ in the social mechanisms that regulate child care and the emotional and sexual development of children.

Cultures, as the before-mentioned matriarchal ones, tend to raise the child within the child's own natural continuum whereas all other cultures tend to condition the child to certain cultural or ideological values and a rigid morality codex.

This is *primarily done through indoctrination* and, secondly, through gradually alienating children from their bodies. The most effective way to indoctrinate children with a certain culture is to implant in their mind a deeply rooted doubt about *who they are*. This doubt which creates a vacuum will then be filled with magic formulas such as *Be not what you are!*

The next step is to force the child to play roles in order to *please their parents*. The main role in this drama which is the *Drama of the Gifted Child*, as Alice Miller called it, is the role of the child parenting his own parents.

> —Alice Miller, The Drama of the Gifted Child: In Search for the True Self (1996).

This is why narcissism is rampant in Western nations, especially in the United States. However, there are few researchers who see that the main etiology for narcissism is to be found in our child rearing paradigm. Those who do, such as Alice Miller or Alexander Lowen are not representing mainstream psychology, despite the brilliance of their work. They have, *inter alia*, found that education that typically leads to narcissism is rich in inventing and executing several other magic formulas that are given to the child in the form of hypnotic injunctions.

> —Alice Miller, Thou Shalt Not Be Aware (1998), and Alexander Lowen, Narcissism: Denial of the True Self (1997).

Some of these are:

- Be adaptable and flexible until self–alienation;
- Never be yourself in front of your parents;
- Be not child-like, but adult-like;
- Be mature in immaturity;
- Understand what your parents don't understand;

# THE SELF-REGULATION PATTERN

- Be logical and uncomplicated;

- Respect your parents while disrespecting yourself;

- Mistrust your intuition;

- Follow authority without questioning.

Many parents who believe they are modern and generous to their children are in reality tyrants because they consume-train their children and mold them into co-dependent *ersatz* partners. In most cases such parents are not conscious about the fact that they act as the long arm of political systems and ideologies subtly hypnotizing their children with the concepts they have themselves been fed with. It is for this reason still considered as revolutionary, if not some sort of subversive activity, to rear children through truthful language and by granting them autonomy, for such kind of education is *not compatible* with the oedipal-paranoid worldview that mainstream industrial culture is founded upon.

To raise children responsibly does not mean to charge them with a burden of premature responsibility. However, to infantilize children and deny them by law any even slight responsibility is surely worse, and the latter practice is the strategy of Oedipal education which arrogantly calls itself *child-protective.* The Oedipal confusion brings about over-adapted and deeply disloyal citizens! Oedipal culture is a community of secret anarchists that obediently say their credo, but silently sabotage the very content of it.

Education toward autonomy is based upon the *unique truth of every single child*, also and especially if this individual truth is contradicting the reigning sociopolitical ideology. It is especially disturbing for the industrial culture that the child be a complete sexual being from birth, and that, as a result, children have a birthright to have their emotions and sexual feelings respected.

Françoise Dolto, in her book *La Cause des Enfants (1985)*, wrote that it scandalizes most adults that a child be their equal and that, therefore, most parents raise their children as formerly princes ruled their kingdoms.

> —Françoise Dolto, La Cause des Enfants (1985), p. 13. See also Ronald David Laing, Divided Self (1991), Bruno Bettelheim, A Good Enough Parent (1988) and The Uses Of Enchantment (1989).

## THE SELF-REGULATION PATTERN

The sociopolitical reasons why this is so are obvious: a body-conscious child is not a consumer of artificial toys and a thousand needs and devices deliberately produced by the industrial consumer culture. For those who object this view, I recall that the repression of the child's sexuality has exactly started with the onset of the Western industrial bourgeoisies, at the end of the 17$^{th}$ century.

> —Françoise Dolto, La Cause des Enfants (1985), pp. 28, 29, citing Ariès, The Childhood of French King Louis XIII.

Françoise Dolto writes:

> Until he was six years old, adults behaved with the prince in a perverse way: they played with his penis, allowed him to play with their genitals and to sleep with them and play *the little devil* with them. All this was allowed. But suddenly when he was six years old, they dressed him like an adult and he had to follow the royal etiquette [citing Ariès, L'Enfant et la vie familiale sous l'Ancien Régime, p 145]. Despite the trauma that could follow, he had nonetheless kept something essential since, during the first years of his life, he could live his sexuality with other adults than his parents. He had here more chance in spite of the precocious adult clothing they put him in. His example is only valid for the rich classes. However, in other levels of society, how could a child of that time repress his incestuous desire and sublimate it? (Id., Translation mine)

Most researchers came to agree that this is the true reason for the repression of child sexuality; while patriarchy has well repressed the sexuality of the girl child long before, this was not true for the boy child. Historical studies on child rearing practices in Europe stress the fact that still during Renaissance the sexuality of the child was not interfered with, and that, back in the Middle-Ages, apart from Christian circles, it was completely free.

The paradigm of *child repression*, as I came to call it, clearly started out with industrial culture, and as a matter of fact, it treats both sexes equally in the denial of child sexuality, and thus cannot be attributed to patriarchy, as some researchers wrongly believe, but is a modern phenomenon, and related to consumerism and materialism.

Consumerist industrialization brought the societal replacement of body pleasure (*to be*) by *ersatz* body pleasure (*to have*), to refer to Erich Fromm's psychoanalytical research on this matter. Ersatz body pleasure is the pleasure that replaces original body pleasure; thus first of all *the toy*. Not the self-made toy that still has some connection with the body, but the industrially produced toy that is completely alien

## THE SELF-REGULATION PATTERN

to the child's body. Typically this toy—which is produced by a gigantic worldwide industry—consists of materials that are not akin to the human body, that is *plastic* and *metal*. Both materials have in common that they are cold and rigid while the body is warm and pliable.

Unconsciously children are conditioned upon the characteristics of the toys they are playing with. *Be plastic!* Be without feelings, artificial. *Be metal!* Be hard and mechanical These are the characteristics of the culture you are growing into. *So mold yourself accordingly!*

Techniques of confusion are all the educational methods that gradually alienate the child from their own truth—which is their body continuum. *The child namely thinks from the body toward the spirit (inductively) while the conditioned adult thinks from the spirit towards the body (deductively).* This means that the child's truth is defined and experienced as the truth of their own body. Every truth that disregards their body or tries to set it aside will not be regarded by the child as truth.

It is for this reason that children cannot comprehend moralistic educational concepts since those concepts deny the body and hypertrophy the intellect. The result are lifelong giant babies in the form of adults who never made the cut with their childhood and remain psychosexually immature: *true virgins*. While life has not made us to remain virgins, but to leave virginity and grow into loving copulation, for otherwise life could not continue.

The fundamental conditioning of man is done within his first seven years of life. What comes later is only polish. *Oedipal confusion cheats about this truth.* It creates a confused spirit within an immature and rigid *nonsexual* body that has lost its natural intelligence.

Oedipal confusion plays the game of eternal maternity until the baby is far older than thirty! It loves naive mother dependence and mistrusts children who are precociously mature. It blows up the child care industry to a gigantic worldwide *business* with children as their products! *Children who resist oedipal castration and maintain their natural capacity for sexual love and sensual pleasure and curiosity are put in the corner and labeled as sexualized and delinquent.* If they still dare to play their own game, the child psychiatrist is ready to interfere and to issue

# THE SELF-REGULATION PATTERN

a certificate which will mark the social death blow: *schizophrenic* or *epileptic*.

*Summerhill School* was founded in 1921 in a village near London, England. It was a *free school* which means that there was no moralistic education and no punishments. What is the difference between the free school concept and other alternative educational concepts? Unlike in concepts such as *Montessori* or *Steiner*, in free schools no repression is used in educating the child. True, Montessori looks very pragmatic, very rational and focused upon the necessities of daily life. Children learn to iron shirts, to do gardening, to cook. They are put in sensitization classes to stimulate their sensorial perception—but this enchanting rhetoric is deeply false. The child does not need to be stimulated sensually since it is naturally sensual. But of course, the moralistic and in this case Catholic background of Montessori education does not allow the children to live out their emotions and display their *sensual and tactile* needs. So they are first emotionally killed and then artificially wrapped into fake-feelings in so-called 'sensitization classes'—a ridiculous and hypocrite procedure after all. With Steiner, it's the stressing of the *soul values* of the child, as if the child by itself was not able to realize their soul level, which is why I came to think that Waldorf education is just another make-believe that is unable to see the reality of children when they are free to develop their natural emotions.

Of course, behind these different approaches are different root philosophies regarding the role of the child in society. Neill, the founder of Summerhill, was against Montessori that he considered as a milder and more intellectual but nonetheless intrinsically authoritarian form of modern education.

There is no doubt that Maria Montessori who was a believing and practicing Christian wanted to revolutionize education and her contribution for more humanity and respect towards children was certainly remarkable. Her teachings brought about amazing change, not only in her own schools. One of her inventions was the child-oriented furniture that we know from modern day care centers. But was it really an approach that served children to be more able to realize their own intrinsic nature?

Montessori's point of departure was the observation that the child's brain, not unlike a sponge, absorbs the intrinsic atmosphere of his or her

# THE SELF-REGULATION PATTERN

environment. In her book *The Absorbent Mind (1973/1995)*, she cites psychological research that proves that children learn in the first three years of their life more than adults in sixty years of hard study.

Although Montessori was in the beginning against all educational programs, she designed a specific educational program for her own schools that focused in first line on the *intellectual training* of the child. By means of a sophisticated system of different games, puzzles and assembler games that by the way are much more complicated than what usually can be bought in toy stores, the child's mind is well prepared to handle both *intellectual and practical* situations in daily life. As such, the theory really makes sense and actually was very appealing to me at first. But what about the soul level of the child and their communication abilities?

My experience of this approach when actually visiting Montessori Schools was rather negative. First it was extremely difficult to get a permission for visit at all. I had to justify my wish as if I wanted to visit a secret terrain of the armed forces. The permission was conditioned upon my being very short and my restraining from any communication with the children.

The children seemed to be robots, pale, dull, insensitive, without life. But they worked, and how! Their way to work though the various tasks seemed obsessive, almost neurotic, while they were bombarded with a full-cry Beethoven symphony from a portable stereo.

These schools really were places without soul and without humor. When the children had their pause, they sat down on a bench, opened their lunch boxes and eat silently, without talking to each other. They seemed to have no contact to each other, isolated in their intellectual cages, without all what usually makes the somewhat noisy and happy society of natural carefree children. These children were little adults that had lost their souls and stared in the air with moody faces. I was highly irritated when I left those places. I found them even worse than the violent orphanages I had seen in some third-world countries. This was child dressage of its best, but one that was no circus because here not even clowns were allowed.

I never got a chance to visit Summerhill. I doubt that it is as permissive as it pretended to be, and this simply because it is located in *England*. Can one

## THE SELF-REGULATION PATTERN

imagine a more repressive culture as to the expression of natural human emotions?

Of course, it is much more humane than most traditional schools and, in particular, *corporal punishment* is absolutely taboo. The education is permissive regarding the healthy emotional and sexual development of the child. Masturbation is not repressed, sexual play only for the purpose to avoid procreation.

That is at least what we learn from Neill's books, but it must seriously be doubted if in practice, the free sexuality of children and youth would ever be tolerated in a school. Today, more than half a century later, it is *not tolerated*. Thus we have to take Neill's statements, *cum grano salis*. Nonetheless, Summerhill has followed up to a great tradition. Neill's educational approach can be seen on a line with famous historical educational methods, like *Jean-Jacques Rousseau*'s educational ideas, or *John Locke*'s, in that it was founded upon nature being generally good.

—Jean–Jacques Rousseau, Émile ou de l'Éducation (1762). See also critically René Schérer, Émile Perverti ou des rapports entre l'éducation et la sexualité (1974) and John

> Locke, Some Thoughts Concerning Education (1690), Vol. IX., pp. 6–205.

Summerhill was thus in flagrant contradiction with Christian, and in particular Calvinistic child rearing which uses harsh punishments and emotional frustrations in order *to better the human soul* that it considers as basically bad and rotten.

The goal of Summerhill was not to bring about conformists, but adults with high self-esteem, strong intuition and sensitivity, humor and respect for life in all its forms.

> —Alexander S. Neill, Summerhill (1961) and Neill! Neill! Orange Peel! (1972).

One of Neill's main motivations was to produce adults with a positive and constructive mindset, people who are free of hatred and the repressed anger that is part of traditional education; people also whose emotional life is intact and lively.

When Neill opened Summerhill, he was already fifty-one years old. He had spent not years but decades with studying human history and education and was deeply concerned that human history was marked by hate and violence, human destructiveness,

## THE SELF-REGULATION PATTERN

intolerance, war and slavery. Neill got to see behind the veil of lies of modern civilization. He saw this perverting hatred again and again in the children who arrived from traditional institutions where they had been declared *uneducable*; he saw that the whole concept of the *difficult child* was wrong in that these children were not more destructive than others, but unhappy, emotionally blocked, frustrated and lonely. Their destructiveness and violence was but a symptom of the underlying reasons that were deeply rooted in the hypocrite, paranoid and violent societal system they were raised in.

—Alexander S. Neill, Summerhill (1961), pp. 139 ff. und pp. 301 ff.

In addition, most of them had been neglected or even abused by their parents or by educators in the homes they were coming from. These children had lost confidence in adults. They had been deceived and felt the cold pressure that comes from authoritarian ways of childrearing. They doubted the existence of love and how it feels to being loved. In this sense, Neill acknowledged, all children who are raised in an authoritarian and repressive system will be difficult once they are freed from the pressure they are subjected to.

This difficult behavior, Neill found, actually was an inner healing process that established a new value system in their mindset. But first they exploded, of course.

With adults it is the same, as we all know. Violent crimes, war, slavery, torture and terrorism are the results of the hypocrite make-believe that we call civilization and that has in truth nothing to do with being civilized. All these consequences of the authoritarian system show that this system is based upon wrong premises, and that it is not human and not made for humans since it disregards human dignity in the most flagrant way.

Neill knew that only an education that was based upon love could finally overcome the violence inherent in a society that is full of hatred. For the only way out of violence is fighting its roots: lack of love and respect, lack of positive encouragement, dehumanizing treatments and a belief system that is based on the idiotic and arrogant idea that nature is fundamentally flawed and has to be improved.

# THE SELF-REGULATION PATTERN

Corroborating Wilhelm Reich's research, Neill found that moralistic education not only negatively impacts upon the mindset of the children, but also infringes upon the soma, especially the energetic and muscular balance in the body of the child. Also in accordance with Wilhelm Reich, Neill applied in his school the principle of *self-regulation*.

Every attempt to impact upon children in a purely intellectual or moralistic way was leading to failure and was soon abandoned by Neill. Instead Neill believed that the children will comply with what they really subscribe to and understand so that they do it because they believe that it has to be done. This attitude of course requires that the educator himself believes that children are beings born with reason and that they will use this reason whatever their age is.

That is exactly where moralistic educators have doubts. They often dig a ravine between adults and children, as if the child lived in another world with different natural laws. While they acknowledge the necessity of reason, they deny that children possess this reason and rather put children on one level with animals. Free childrearing is unthinkable without conceding children the freedom of their natural erotic and sexual feelings and, not to forget it, the freedom of speech regarding these feelings.

> —See also Floyd M. Martinson, The Sex Education of Young Children (1981), pp. 51 ff.

The latter is as important as the former because children who are allowed to do it without being allowed to talk about what they have done will never really believe that this freedom has been given to them. Instead they will experience guilt and shame. Only a child who is sexually free and erotically satiated will develop their full potential of interest and work energy. It is further necessary that children really *feel accepted* as persons in their own right, in their natural wholeness that encompasses an undistorted emonic setup that wants to be developed and experienced, for all life asks for growth, and it is therefore total nonsense when mainstream psychologists pretend the child had to grow with a *sleeping* sexuality that would *awake* at the end of puberty. Fairy tales.

We are facing this challenge for a more truthful education also on a collective and global level when we develop more tolerance, understanding and compassion for others. It is by changing the basic

## THE SELF-REGULATION PATTERN

foundation of our educational and pedagogical values. The Summerhill concept realized a first step to more humanity here on earth, by raising more *humane humans*. Interestingly, the overwhelming majority of Summerhill graduates were later seen to score very well on the social ladder while leading healthy and balanced lives and experiencing positive and highly rewarding relationships. Neill stated that his own measure of success was the capability to work with joy and to live positively.

—See A. S. Neill, Summerhill (1961), p. 29.

After forty years of experience with Summerhill, Neill was able to conclude that, applying this definition of success, *'most of the Summerhill graduates became successful people.'* (Id.)

# The Touch Pattern

In our observation of the *Love Pattern* and *Pleasure Pattern* and of the nonviolent and sexually permissive Trobriand culture, we saw that originally love and sex were not separated, but were linked in a functional way. For developing our love capacity, we need to live as children our sexual curiosity to its fullest. In order to be able to remain with a partner over many years, we need to have had promiscuous sex previously, and preferably during childhood.

Instead, what we can derive from the anthropological field research is that our Western theory of love and sex being *separate* experiences is fundamentally schizoid, and contradicts the nonlinear logic of life and all insights into the connectivity of mind and soma. It seems to be an *elaboration of pornographic clerical fantasies* rather than the worldview of sane and integrated individuals. The word *sex* itself, that interestingly is missing in the vocabulary of many a sexually permissive culture, implies, by the very fact that such word exists, the possibility of a split between sexual function from love so that love remains a kind of reductionist concept of 'pure caring and affection' which, of course, does not exist in real life.

*This split between love and sex makes for a lot of damage in our striving for unity and harmony.* One of the typical problems that many Westerners face is that they unconsciously split off their love feelings and their sexual longings and project them on different partners. The marriage partner receives love, care and affection and all sexual longings are split off and projected upon prostitutes or prostitute-ersatz figures such as secretaries, servants or female children.

Many a reader may object that it often happens that sexuality is lived in its cold form, deprived of love, as mere satisfaction and rather brutally, and that some people even experience more intense sexual feelings when they can encounter sex without being obliged to fake tenderness or caring. This may be true but research has also shown that sex which is experienced connected with love engenders a higher level of lasting feelings of happiness and joy than sex that is acted out in form of an ego trip.

# THE TOUCH PATTERN

Whatever our personal opinion in this respect may be like, there is no doubt about the fact that *sexuality in every form is focused upon our skin as the main sexual organ*. It seems, however, that sexology has rather hesitantly taken the turn to consider the importance of skin contact in the give and take of body pleasure.

Without the skin we would be like barrels without ground. Our bodies are water-sleeves containing more than ninety percent of water. Besides, the skin is our temperature regulator, similar to the astronaut's spacesuit. Besides this protective function, our skin plays an important role in our wellbeing and health. When we experience to be caressed or massaged, when we are touched with love, we feel good. *It feels good.*

Ashley Montagu spent more than thirty years researching about the overwhelming importance of children's tactile stimulation. His book *Touching: The Human Significance of the Skin (1971)* is a guide not only for his own research results, but also for the complete range of literature on skin research collected over twenty years, in this study. Ashley Montagu gives many examples that demonstrate how destructive the *touch taboo* in our culture is, especially for children. He found that the deprivation of tactile pleasure creates a misbalance in the psychosomatic setup of the child.

As I reported it already, Montagu, starting out with animal research, inquired about the biological reasons for the motherly licking as a primary condition for the survival of the young. It is noteworthy to observe that the mammal mother extensively licks the perineal zone, the region between anus and genitals of the young. Experiments that prohibited test animals from licking and being licked led to grave urinogenital infections of the young animals or even their death.

Further, Montagu found that licking is normally *not* part of human child care. An exception was found with an Eskimo tribe, the *Ingalik*. The Ingalik mother would lick the face and the hands of the newborn in order to clean them. She would continue this licking until the baby is old enough to sit on the bench.

Montagu found that humans generally touch children more with their hands, that is by touching, stroking, caressing, than, for example, with their tongues, and thus by licking, and that, in addition, eye-to-eye contact plays an important role in nurturant child care.

Different researchers found that tactile stimulation of the child is of primary importance for building and maintaining a strong immune system with the child's organism.

—Id., pp. 20, 21 and Michel Odent, Birth Reborn (1986).

Montagu remarks in this context that love has been defined as *the harmony of two souls and the contact of two epidermises.* In this sense body pleasure of mother and child is the most basic, the most natural and the most beneficial form of human sexuality.

Needless to add that this is a form of *pedophile* sexuality. And this is one of the examples that clearly show that pedophile erotic and sexual sensations are in some ways part of nature, and not per se a perversion.

The skin is our primary sexual organ. All stimulation of the sexual organs is effected through the stimulation of the skin that surrounds them. Sigmund Freud has defined sexuality as *any behavior that has a physical connotation to the sexual organs and that is focused on receiving pleasure.* We could ask if body pleasure really must be concentrated on the sexual organs so that we can qualify it as sexual? It is certainly also a form of body pleasure to drink a fresh beer or to eat one's favorite dish.

This kind of pleasure could be called oral or nutritive pleasure. Hardly anyone would go as far as qualifying it as *sexual*. But how is it with caressing one's chest or bottom? Is it sexual or not? Does it depend on the way we caress that it is sexual or merely affectionate, or does it depend on the intention? Or is the decisive factor which body zone is caressed? Or does it depend on the fact that the one who caresses is sexually aroused by the activity—or not?

Still during the Renaissance it was common in Europe that all members of the family slept naked in one bed, as today it still is practiced with the Eskimo and many native populations. The bodily touch or casual caresses that happened during the night were generally *not* considered as sexual or sexually intended. Today, in our culture, many people would find it unusual to let sleep their children naked in one bed or that adults, be it the parents, would sleep naked together with their children. This is astonishing as the majority of psychologists are now outspoken about our need for direct body contact, warmth, togetherness, tenderness, and nudity—and this independently of age or gender. Several scientists have researched on the consequences of a deprivation of love nutrition, that is the lack of tactile pleasure, and got alarming results. Unfortunately, the greatest part of this research was done with apes, although genetically, as we all know, the human race is very close to apes. But nonetheless it would seem more convincing to have achieved these same results with direct research on human beings.

Pediatrics and modern child psychology have done great work during the last twenty years. After the publication of volumes if not entire libraries of results of this research, now almost all specialists of child care agree that *children who are raised in a milieu deprived of love, tenderness and caring body touch face greater adaptation problems later in life*, frequently show learning difficulties and tend to be more rigid in experiencing joy and pleasure than children who grew up with love and body touch.

The first group of children exhibit symptoms such as restlessness or hyperactivity; in school they often have drawbacks because of their low attention span and concentration ability. In the group, they are seen as rather isolationist and uncooperative. They are easily pushed aside as *difficult*, and once this has happened, the above symptoms aggravate, sometimes dramatically.

Many of the children who are in institutions for so-called delinquent youth were formerly affectively and tactually deprived children; yet life circumstances and often an intolerant and punitive attitude from the side of the environment made them turn away from sociability and into marginality.

What is the specifically pathological in their behavior and in the circumstances that have contributed to form it? How does it impact on children if in their family tenderness and care were replaced by violence and brutality? Research into domestic violence has shown that healthy forms of body touch and body pleasure do virtually not exist in such families. If there is touch at all, it is one that hurts, violates, humiliates and degrades. There is almost unanimity among scientists and professionals that for the small child *tactile stimulation is essential* for healthy psychosomatic growth. It has been shown that close and long-term body contact between the child and their mother, father or other tactually nutritive persons decisively strengthens the child's immune system and improves health. One could conclude that these findings are not only valid for small children but also children between the age of six until puberty, and even adolescents, for what could be called *skin erotic* seems to be a life-enhancing and health-strengthening factor in all living.

In fact, in India, as Frederick Leboyer reports, where it is a common tradition to massage babies with warm oil, there are many mothers who continue massaging their children, which always includes gently massaging their genitals, until adolescence.

—Frederick Leboyer, Loving Hands (1977).

It is believed that massaging children's genitals will enhance their procreative ability, sexual potency and resistance against illness. Such tactile forms of childcare are however by no means associated in India with incest or pedophilia nor is it even considered as sexual. It is regarded as a natural and necessary attribute to nurturant parental care.

Much research has been done on the roots of human violence, but there is hardly anything comparable to the findings of the American sociologist *James W. Prescott*. His publications are unique in that they found child care methods, sexual attitudes and violence level in a given society forming part of a subtle feedback system.

>—James W. Prescott, Body Pleasure and the Origins of Violence (1975).

As already pointed out, the quintessence of this research is the thesis that cultures that continue to be highly repressive regarding the emotional and tactile needs of small children and that, in addition, prohibit premarital sex, will end in an inevitable chaos of violence and destruction that has not seen an equal in human history.

Violence, Prescott states authoritatively, and with abundant amount of proof, is a *compensation reaction* of the human brain for the deprivation of tactile pleasure!

These research results amazingly match with the findings of the British neurophysiologist Herbert James Campbell, who summarized forty years of neurological research saying that the motivation of every kind of human activity is nothing else but pleasure.

—Herbert James Campbell, The Pleasure Areas (1973).

In case pleasure is prohibited, the human brain automatically compensates for this lack by activating the violence areas in the brain. The pleasure areas and the violence areas function, neurologically speaking, in a way that one inhibits the other; their activity levels are thus mutually exclusive. For example, the more the pleasure areas are stimulated, the more the violence areas will be inhibited and deactivated. The other way around is equally true: the more the human brain runs on violence, the more its capacity of experiencing pleasure is infringed.

Thus, to summarize Prescott's findings provocatively, one could say that it is up to us, and within our individual responsibility, if we want to run our brains on violence or on pleasure. Both is *not* possible. We have to decide what we want. American society seems to have made this decision long ago, Prescott found. It has unequivocally decided for violence and against pleasure. The historical roots?

Prescott identifies them in the Biblical and Jewish traditions that are the foundation of American or, more generally, Western society. The logical conclusions of this interconnectedness are:

- the more a person has received tactile nutrition during her early years, the more she has known body pleasure from childhood, the more her pleasure areas will be activated and, as a result, the more her violence areas will be inhibited;

- the more a person was deprived of tactile pleasure during childhood, the more her desire for body pleasure was repressed or body pleasure experienced as a guilt-producing activity, the more the person's pleasure areas will be inhibited and, as a result, the more her violence areas will be active.

Interestingly, Prescott found that tactile deprivation in early childhood does not automatically lead to a violent character. Interestingly, there are factors that compensate for early tactile deprivation, the most decisive of those factors being pre-marital sex.

—James W. Prescott, Body Pleasure and the Origins of Violence (1975), p. 13.

## THE TOUCH PATTERN

In this point resides the specific appeal for a future world politics that is peace-centered. This appeal is to refrain from repressing children's emotions, their sensuality, and their budding sexuality, to let it grow freely and without moralistic or other inhibitions, and to recognize this freedom by law.

Unfortunately Prescott's research *never really penetrated into the mass media* that seem to discard out from public forums this kind of research and is known only to a relatively small circle of scientists and intellectuals. The problem of the present public discussion about sex is that it is still rigid and impregnated with various fears and taboos. The *stranger* haunts American talk-shows and the general hysteria about abusive or non-abusive sex with children is all-pervasive in the American media culture.

This general attitude is far from being supportive for the tactile needs of present-day Western children. Its obsessive focus on protecting children can easily be revealed as either a lip-service, or a new money-making device, or else another way to enslave children into a doctrinaire system of industrial or post-industrial values.

To partly remedy the present almost hopeless situation would be, for example, to implement *baby massage* on a large scale as it was proposed by Frederick Leboyer. This would implement at least one possible form of tactile stimulation and body pleasure for the growing next generations. It would ensure that next generation societies will be a lot more peaceful and a lot less violent than the ones preceding them in history.

We spontaneously communicate love through body touch and skin contact. Parents fondle and kiss their baby. Lovers embrace each other. Children like to cuddle into their parents' bed and siblings naturally share one bed, at least until a certain age. Attitudes in this respect vary from one society and continent to the other.

Historically body touch was natural and not reflected upon. Sexual contacts in certain institutionalized forms are to be traced back into the beginnings of human life and history.

—See for example Susanne Cho, Kindheit und Sexualität im Wandel der Kulturgeschichte (1983) or Floyd M. Martinson, The Sex Education of Young Children (1981), pp. 51 ff.

# THE TOUCH PATTERN

Moreover, as the French child therapist and book author Françoise Dolto reports in her book *La Cause des Enfants (1985)*, still in the seventeenth century love and sex games between adult women and small boys were not repressed and happened quite often. They may have been considered by some people as funny or scurrilous, but not as immoral and still less as dangerous.

> —This fact is corroborated by descriptions of the habits of the Royal Family in France, as reported by the doctor of Louis XIII, Héroard, see: J. Héroard, Journal de Jean Héroard sur l'Enfance et la Jeunesse de Louis XIII (1868). See also: Lloyd deMause (Ed.), The History of Childhood (1974), p. 23 and Philippe Ariès, Centuries of Childhood (1962).

The separation of love and eroticism into different age groups and the sexual mathematics that results from this modern kind of regimenting sexual relations, has brought about much confusion and, as a result, much public discussion.

After all, it seems to be entirely unnatural and artificial, to say the least. In fact, a generally schizoid and punitive attitude toward love, sex and tenderness can be traced back to the Assyrians and, in particular, the *Code of Hammurabi*, the first legal code in history that contained sex laws. In addition, the so–called holy books of all our important religions are filled with poisonous negativism regarding body pleasure.

Tolerance, and as an essential ingredient sexual tolerance, has never been practiced by humanity. As a matter of fact, historically, the times of repression and persecution of sexual minorities by far prevail in human history. After all, modern science cannot compensate for the darkness and stubborn rigidity of a human mind that has been conditioned to violence since centuries.

Only love can bring a change. The problem is not that people are not enough informed or not interested in science, but that most individuals are relying too much upon moral judgments that they have more or less blindly taken over from higher authorities, instead of relying on their own intuitions and the inherent intelligence of their bodies and their skins. It does not make sense to start witch hunts in order to eradicate witch hunts. It is more intelligent to leave those how are armored in their armor and to take care of those who have not yet built such armor against life and against pleasure. Our skin can show us the way and gives the signals.

## THE TOUCH PATTERN

In order to find back to our natural continuum, we must relearn to reason in a way that includes touching and feeling and to allow, in a real and a metaphoric sense, our skin to remain open for the osmosis of love.

# BIBLIOGRAPHY

### ABRAMS, JEREMIAH (ED.)

RECLAIMING THE INNER CHILD
New York: Tarcher/Putnam, 1990

### ALSTON, JOHN P. / TUCKER, FRANCIS

THE MYTH OF SEXUAL PERMISSIVENESS
The Journal of Sex Research, 9/1 (1973)

### APPLETON, MATTHEW

A FREE RANGE CHILDHOOD
Self-Regulation at Summerhill School
Foundation for Educational Renewal, 2000

### ARCAS, GÉRALD, DR

GUÉRIR LE CORPS PAR L'HYPNOSE ET L'AUTO-HYPNOSE
Paris: Sand, 1997

### ARIÈS, PHILIPPE

L'ENFANT ET LA FAMILLE SOUS L'ANCIEN RÉGIME
Paris, Seuil, 1975

CENTURIES OF CHILDHOOD
New York: Vintage Books, 1962

GESCHICHTE DER KINDHEIT
Frankfurt/M: DTV, 1998

## ARNTZ, WILLIAM & CHASSE, BETSY

WHAT THE BLEEP DO WE KNOW
20th Century Fox, 2005 (DVD)

DOWN THE RABBIT HOLE QUANTUM EDITION
20th Century Fox, 2006 (3 DVD Set)

RELATIONSHIPS AND LIFE CYCLES
Astrological Patterns of Personal Experience
Sebastopol, CA: CRCS Publications, 1993

## ATLEE, TOM

THE TAO OF DEMOCRACY
Using Co-Intelligence to Create a World That Works for All
North Charleston, SC: Imprint Books / WorldWorks Press, 2003

## BACHELARD, GASTON

THE POETICS OF REVERIE
Translated by Daniel Russell
Boston: Beacon Press, 1971

# BIBLIOGRAPHY

## BAGGINS, DAVID SADOFSKY

DRUG HATE AND THE CORRUPTION OF AMERICAN JUSTICE
Santa Barbara: Praeger, 1998

## BAGLEY, CHRISTOPHER

CHILD ABUSERS
Research and Treatment
New York: Universal Publishers, 2003

## BALTER, MICHAEL

THE GODDESS AND THE BULL
Catalhoyuk, An Archaeological Journey
to the Dawn of Civilization
New York: Free Press, 2006

## BANDLER, RICHARD

GET THE LIFE YOU WANT
The Secrets to Quick and Lasting Life Change
With Neuro-Linguistic Programming
Deerfield Beach, Fl: HCI, 2008

## BARBAREE, HOWARD E. & MARSHALL, WILLIAM L. (EDS.)

THE JUVENILE SEX OFFENDER
Second Edition
New York: Guilford Press, 2008

## Barron, Frank X., Montuori, et al. (Eds.)

Creators on Creating
Awakening and Cultivating the Imaginative Mind
(New Consciousness Reader)
New York: P. Tarcher/Putnam, 1997

## Bateson, Gregory

Steps to an Ecology of Mind
Chicago: University of Chicago Press, 2000
Originally published in 1972

## Bender Lauretta & Blau, Abram

The Reaction of Children to Sexual Relations with Adults
American J. Orthopsychiatry 7 (1937), 500-518

## Bernard, Frits

Paedophilia
A Factual Report
Amsterdam: Enclave, 1985

## Bertalanffy, Ludwig von

General Systems Theory
Foundations, Development, Applications
New York: George Brazilier Publishing, 1976

# BIBLIOGRAPHY

## BESANT, ANNIE

AN AUTOBIOGRAPHY
New Delhi: Penguin Books, 2005
Originally published in 1893

## BETTELHEIM, BRUNO

A GOOD ENOUGH PARENT
New York: A. Knopf, 1987

THE USES OF ENCHANTMENT
New York: Vintage Books, 1989

## BOHM, DAVID

WHOLENESS AND THE IMPLICATE ORDER
London: Routledge, 2002

THOUGHT AS A SYSTEM
London: Routledge, 1994

QUANTUM THEORY
London: Dover Publications, 1989

## BOLDT, LAURENCE G.

ZEN AND THE ART OF MAKING A LIVING
A Practical Guide to Creative Career Design
New York: Penguin Arkana, 1993

HOW TO FIND THE WORK YOU LOVE
New York: Penguin Arkana, 1996

ZEN SOUP
Tasty Morsels of Zen Wisdom From Great Minds East & West
New York: Penguin Arkana, 1997

THE TAO OF ABUNDANCE
Eight Ancient Principles For Abundant Living
New York: Penguin Arkana, 1999

## BORDEAUX-SZEKELY, EDMOND

TEACHING OF THE ESSENES FROM ENOCH TO THE DEAD
Sea Scrolls
Beekman Publishing, 1992

GOSPEL OF THE ESSENES
The Unknown Books of the Essenes
& Lost Scrolls of the Essene Brotherhood
Beekman Publishing, 1988

GOSPEL OF PEACE OF JESUS CHRIST
Beekman Publishing, 1994

GOSPEL OF PEACE, 2D VOL.
I B S International Publishers

## BRANDEN, NATHANIEL

HOW TO RAISE YOUR SELF-ESTEEM
New York: Bantam, 1987

# BIBLIOGRAPHY

## Brant & Tisza

The Sexually Misused Child
American J. Orthopsychiatry, 47(1)(1977)

## Bullough & Bullough (Eds.)

Human Sexuality
An Encyclopedia
New York: Garland Publishing, 1994

Sin, Sickness and Sanity
A History of Sexual Attitudes
New York: New American Library, 1977

## Buxton, Richard

The Complete World of Greek Mythology
London: Thames & Hudson, 2007

## Cain, Chelsea & Moon Unit Zappa

Wild Child
New York: Seal Press (Feminist Publishing), 1999

## Calderone & Ramey

Talking With Your Child About Sex
New York: Random House, 1982

## Campbell, Herbert James

The Pleasure Areas
London: Eyre Methuen Ltd., 1973

## Campbell, Jacqueline C.

Assessing Dangerousness
Violence by Sexual Offenders, Batterers and Child Abusers
New York: Sage Publications, 2004

## Campbell, Joseph

The Hero With A Thousand Faces
Princeton: Princeton University Press, 1973
(Bollingen Series XVII)
London: Orion Books, 1999

Occidental Mythology
Princeton: Princeton University Press, 1973
(Bollingen Series XVII)
New York: Penguin Arkana, 1991

The Masks of God
Oriental Mythology
New York: Penguin Arkana, 1992
Originally published 1962

The Power of Myth
With Bill Moyers
ed. by Sue Flowers
New York: Anchor Books, 1988

# BIBLIOGRAPHY

## CAPACCHIONE, LUCIA

**THE POWER OF YOUR OTHER HAND**
North Hollywood, CA: Newcastle Publishing, 1988

## CAPRA, BERNT AMADEUS

**MINDWALK**
A Film for Passionate Thinkers
Based Upon Fritjof Capra's The Turning Point
New York: Triton Pictures, 1990

## CAPRA, FRITJOF

**THE TURNING POINT**
Science, Society And The Rising Culture
New York: Simon & Schuster, 1987
Original Author Copyright, 1982

**THE TAO OF PHYSICS**
An Exploration of the Parallels Between Modern
Physics and Eastern Mysticism
New York: Shambhala Publications, 2000
(New Edition) Originally published in 1975

**THE WEB OF LIFE**
A New Scientific Understanding of Living Systems
New York: Doubleday, 1997
Author Copyright 1996

**THE HIDDEN CONNECTIONS**
New York: Doubleday, 2002

STEERING BUSINESS TOWARD SUSTAINABILITY
New York: United Nations University Press, 1995

UNCOMMON WISDOM
Conversations with Remarkable People
New York: Bantam, 1989

THE SCIENCE OF LEONARDO
Inside the Mind of the Great Genius of the Renaissance
New York: Anchor Books, 2008
New York: Bantam Doubleday, 2007 (First Publishing)

## CASTANEDA, CARLOS

THE TEACHINGS OF DON JUAN
A Yaqui Way of Knowledge
Washington: Square Press, 1985

JOURNEY TO IXTLAN
Washington: Square Press: 1991

TALES OF POWER
Washington: Square Press, 1991

THE SECOND RING OF POWER
Washington: Square Press, 1991

## CLARKE-STEWARD, S., FRIEDMAN, S. & KOCH, J.

CHILD DEVELOPMENT, A TOPICAL APPROACH
London: John Wiley, 1986

# BIBLIOGRAPHY

## CONSTANTINE, LARRY L.

CHILDREN & SEX
New Findings, New Perspectives
Larry L. Constantine & Floyd M. Martinson (Eds.)
Boston: Little, Brown & Company, 1981

TREASURES OF THE ISLAND
Children in Alternative Lifestyles
Beverly Hills: Sage Publications, 1976

WHERE ARE THE KIDS?
in: Libby & Whitehurst (ed.)
Marriage and Alternatives
Glenview: Scott Foresman, 1977

OPEN FAMILY
A Lifestyle for Kids and other People
26 FAMILY COORDINATOR 113-130 (1977)

## COOK, M. & HOWELLS, K. (EDS.)

ADULT SEXUAL INTEREST IN CHILDREN
Academic Press, London, 1980

## COVITZ, JOEL

EMOTIONAL CHILD ABUSE
The Family Curse
Boston: Sigo Press, 1986

## CURRIER, RICHARD L.

JUVENILE SEXUALITY IN GLOBAL PERSPECTIVE
in : Children & Sex, New Findings, New Perspectives
Larry L. Constantine & Floyd M. Martinson (Eds.)
Boston: Little, Brown & Company, 1981

## DE BONO, EDWARD

THE USE OF LATERAL THINKING
New York: Penguin, 1967

THE MECHANISM OF MIND
New York: Penguin, 1969

SUR/PETITION
London: HarperCollins, 1993

TACTICS
London: HarperCollins, 1993
First published in 1985

SERIOUS CREATIVITY
Using the Power of Lateral Thinking to Create New Ideas
London: HarperCollins, 1996

## DELACOUR, JEAN-BAPTISTE

GLIMPSES OF THE BEYOND
New York: Bantam Dell, 1975

## DeMause, Lloyd

The History of Childhood
New York, 1974

Foundations of Psychohistory
New York: Creative Roots, 1982

## Diamond, Stephen A., May, Rollo

Anger, Madness, and the Daimonic
The Psychological Genesis of Violence, Evil and Creativity
New York: State University of New York Press, 1999

## DiCarlo, Russell E. (Ed.)

Towards A New World View
Conversations at the Leading Edge
Erie, PA: Epic Publishing, 1996

## Dolto, Françoise

La Cause des Enfants
Paris: Laffont, 1985

Psychanalyse et Pédiatrie
Paris: Seuil, 1971

Séminaire de Psychanalyse d'Enfants, 1
Paris: Seuil, 1982

Séminaire de Psychanalyse d'Enfants, 2
Paris: Seuil, 1985

SÉMINAIRE DE PSYCHANALYSE D'ENFANTS, 3
Paris: Seuil, 1988

L'ÉVANGILE AU RISQUE DE LA PSYCHANALYSE
Paris: Seuil, 1980

## DÜRCKHEIM, KARLFRIED GRAF

HARA: THE VITAL CENTER OF MAN
Rochester: Inner Traditions, 2004

ZEN AND US
New York: Penguin Arkana 1991

THE CALL FOR THE MASTER
New York: Penguin Books, 1993

ABSOLUTE LIVING
The Otherworldly in the World and the Path to Maturity
New York: Penguin Arkana, 1992

THE WAY OF TRANSFORMATION
Daily Life as a Spiritual Exercise
London: Allen & Unwin, 1988

THE JAPANESE CULT OF TRANQUILITY
London: Rider, 1960

## EDMUNDS, FRANCIS

AN INTRODUCTION TO ANTHROPOSOPHY
Rudolf Steiner's Worldview
London: Rudolf Steiner Press, 2005

# BIBLIOGRAPHY

## EDWARDES, A.

THE JEWEL OF THE LOTUS
New York, 1959

## EINSTEIN, ALBERT

THE WORLD AS I SEE IT
New York: Citadel Press, 1993

OUT OF MY LATER YEARS
New York: Outlet, 1993

IDEAS AND OPINIONS
New York: Bonanza Books, 1988

ALBERT EINSTEIN NOTEBOOK
London: Dover Publications, 1989

## EISLER, RIANE

THE CHALICE AND THE BLADE
Our history, Our future
San Francisco: Harper & Row, 1995

SACRED PLEASURE: SEX, MYTH AND THE POLITICS OF THE BODY
New Paths to Power and Love
San Francisco: Harper & Row, 1996

THE PARTNERSHIP WAY
New Tools for Living and Learning
With David Loye
Brandon, VT: Holistic Education Press, 1998

## ELWIN, V.

THE MURIA AND THEIR GHOTUL
Bombay: Oxford University Press, 1947

THE SECRET LIFE OF WATER
New York: Atria Books, 2005

## ERICKSON, MILTON H.

MY VOICE WILL GO WITH YOU
The Teaching Tales of Milton H. Erickson
by Sidney Rosen (Ed.)
New York: Norton & Co., 1991

COMPLETE WORKS 1.0, CD-ROM
New York: Milton H. Erickson Foundation, 2001

## ERIKSON, ERIK H.

CHILDHOOD AND SOCIETY
New York: Norton, 1993
First published in 1950

## EVANS-WENTZ, WALTER YEELING

THE FAIRY FAITH IN CELTIC COUNTRIES
London: Frowde, 1911
Republished by Dover Publications
(Minneola, New York), 2002

BIBLIOGRAPHY

## Farson, Richard

Birthrights
A Bill of Rights for Children
Macmillan, New York, 1974

## Feinberg, Joel

Harmless Wrongdoing
The Moral Limits of the Criminal Law, Vol. 4
New York: Oxford University Press, 1990

## Fensterhalm, Herbert

Don't Say Yes When You Want to Say No
With Jean Bear
New York: Dell, 1980

## Finkelhor, David

Sexually Victimized Children
New York: Free Press, 1981

## Finkelstein, Haim N. (Ed.)

The Collected Writings of Salvador Dali
Cambridge: Cambridge University Press, 1998

## Fortune, Mary M.

Sexual Violence
New York: Pilgrim Press, 1994

## Foster/Freed

A Bill of Rights for Children
6 FAMILY LAW QUARTERLY 343 (1972)

## Foucault, Michel

The History of Sexuality, Vol. I : The Will to Knowledge
London: Penguin, 1998
First published in 1976

The History of Sexuality, Vol. II : The Use of Pleasure
London: Penguin, 1998
First published in 1984

The History of Sexuality, Vol. III : The Care of Self
London: Penguin, 1998
First published in 1984

## Freud, Sigmund

Three Essays on the Theory of Sexuality
in: The Standard Edition of the Complete Psychological Works of Sigmund Freud
London: Hogarth Press, 1953-54
Vol. 7, pp. 130 ff
(first published in 1905)

# BIBLIOGRAPHY

THE INTERPRETATION OF DREAMS
New York: Avon, Reissue Edition, 1980
and in: The Standard Edition of the Complete Psychological Works of Sigmund Freud, (24 Volumes) ed. by James Strachey
New York: W. W. Norton & Company, 1976

TOTEM AND TABOO
New York: Routledge, 1999
Originally published in 1913

## FREUND, KURT

ASSESSMENT OF PEDOPHILIA
in: Cook, M. and Howells, K. (eds.)
Adult Sexual Interest in Children
Academic Press, London, 1980

## FROMM, ERICH

THE ANATOMY OF HUMAN DESTRUCTIVENESS
New York: Owl Book, 1992
Originally published in 1973

ESCAPE FROM FREEDOM
New York: Owl Books, 1994
Originally published in 1941

TO HAVE OR TO BE
New York: Continuum International Publishing, 1996
Originally published in 1976

THE ART OF LOVING
New York: HarperPerennial, 2000
Originally published in 1956

## Geldard, Richard

Remembering Heraclitus
New York: Lindisfarne Books, 2000

## Gerber, Richard

A Practical Guide to Vibrational Medicine
Energy Healing and Spiritual Transformation
New York: Harper & Collins, 2001

## Geller, Uri

The Mindpower Kit
Includes Book, Audiotape, Quartz Crystal And Meditation Circle
New York: Penguin, 1996

## Gesell, Izzy

Playing Along
37 Group Learning Activities Borrowed from Improvisational Theater
Whole Person Associates, 1997

## Ghiselin, Brewster (Ed.)

The Creative Process
Reflections on Invention in the Arts and Sciences
Berkeley: University of California Press, 1985
First published in 1952

# BIBLIOGRAPHY

## GIBSON, IAN

THE SHAMEFUL LIFE OF SALVADOR DALI
New York: Norton, 1998

## GIL, DAVID G.

SOCIETAL VIOLENCE AND VIOLENCE IN FAMILIES
in: David G. Gil, Child Abuse and Violence
New York: Ams Press, 1928

## GIMBUTAS, MARIJA

THE LANGUAGE OF THE GODDESS
London: Thames & Hudson, 2001

## GOLDENSTEIN, JOYCE

EINSTEIN: PHYSICIST AND GENIUS
(Great Minds of Science)
New York: Enslow Publishers, 1995

## GOLDMAN, JONATHAN & GOLDMAN, ANDI

TANTRA OF SOUND
Frequencies of Healing
Charlottesville: Hampton Roads, 2005

HEALING SOUNDS
The Power of Harmonies
Rochester: Healing Arts Press, 2002

HEALING SOUNDS
Principles of Sound Healing
DVD, 90 min.
Sacred Mysteries, 2004

## GOLDSTEIN, JEFFREY H.

AGGRESSION AND CRIMES OF VIOLENCE
New York, 1975

## GOLEMAN, DANIEL

EMOTIONAL INTELLIGENCE
New York, Bantam Books, 1995

## GORDON, ROSEMARY

PEDOPHILIA: NORMAL AND ABNORMAL
in: Kraemer, The Forbidden Love
London, 1976

## GORDON WASSON, R.

THE ROAD TO ELEUSIS
Unveiling the Secret of the Mysteries
With Albert Hofmann, Huston Smith, Carl Ruck and Peter Webster
Berkeley, CA: North Atlantic Books, 2008

# BIBLIOGRAPHY

## Goswami, Amit

The Self-Aware Universe
How Consciousness Creates the Material World
New York: Tarcher/Putnam, 1995

## Gottlieb, Adam

Peyote and Other Psychoactive Cacti
Ronin Publishing, 2nd edition, 1997

## Grof, Stanislav

Ancient Wisdom and Modern Science
New York: State University of New York Press, 1984

Beyond the Brain
Birth, Death and Transcendence in Psychotherapy
New York: State University of New York, 1985

LSD: Doorway to the Numinous
The Groundbreaking Psychedelic Research into Realms of the Human Unconscious
Rochester: Park Street Press, 2009

Realms of the Human Unconscious
Observations from LSD Research
New York: E.P. Dutton, 1976

The Cosmic Game
Explorations of the Frontiers of Human Consciousness
New York: State University of New York Press, 1998

THE HOLOTROPIC MIND
The Three Levels of Human Consciousness
With Hal Zina Bennett
New York: HarperCollins, 1993

WHEN THE IMPOSSIBLE HAPPENS
Adventures in Non-Ordinary Reality
Louisville, CO: Sounds True, 2005

## HOUSTON, JEAN

THE POSSIBLE HUMAN
A Course in Enhancing Your Physical, Mental, and Creative Abilities
New York: Jeremy P. Tarcher/Putnam, 1982

## HOWELLS, KEVIN

ADULT SEXUAL INTEREST IN CHILDREN
Considerations Relevant to Theories of Aetiology in:
Cook, M. and Howells, K. (eds.): Adult Sexual Interest in Children
Academic Press, London, 1980

## HUNT, VALERIE

INFINITE MIND
Science of the Human Vibrations of Consciousness
Malibu, CA: Malibu Publishing, 2000

## INNOCENTI DECLARATION

DECLARATION ON THE PROTECTION, PROMOTION AND SUPPORT OF BREASTFEEDING
http://www.innocenti15.net/inno.htm

## JACKSON, NIGEL

THE RUNE MYSTERIES
With Silver RavenWolf
St. Paul, Minn.: Llewellyn Publications, 2000

## JACKSON, STEVI

CHILDHOOD AND SEXUALITY
New York: Blackwell, 1982

## JAFFE, HANS L.C.

PICASSO
New York: Abradale Press, 1996

## JAMES, WILLIAM

WRITINGS 1902-1910
The Varieties of Religious Experience / Pragmatism / A Pluralistic Universe / The Meaning of Truth / Some Problems of Philosophy / Essays
New York: Library of America, 1988

## JANOV, ARTHUR

PRIMAL MAN
The New Consciousness
New York: Crowell, 1975

## JOHNSON, PAUL

A HISTORY OF THE JEWS
New York: Harper & Row, 1987

## JOHNSTON & DEISHER

CONTEMPORARY COMMUNAL CHILD REARING: A FIRST ANALYSIS
52 PEDIATRICS 319 (1973)

## JONES, W.H.S., LITT, D.

PLINY NATURAL HISTORY
Cambridge, Mass.: Harvard University Press, 1980

## JUNG, CARL GUSTAV

ARCHETYPES OF THE COLLECTIVE UNCONSCIOUS
in: The Basic Writings of C.G. Jung
New York: The Modern Library, 1959, 358-407

COLLECTED WORKS
New York, 1959

ON THE NATURE OF THE PSYCHE
in: The Basic Writings of C.G. Jung

# BIBLIOGRAPHY

New York: The Modern Library, 1959, 47-133

PSYCHOLOGICAL TYPES
Collected Writings, Vol. 6
Princeton: Princeton University Press, 1971

PSYCHOLOGY AND RELIGION
in: The Basic Writings of C.G. Jung
New York: The Modern Library, 1959, 582-655

RELIGIOUS AND PSYCHOLOGICAL PROBLEMS OF ALCHEMY
in: The Basic Writings of C.G. Jung
New York: The Modern Library, 1959, 537-581

SYMBOL UND LIBIDO
Freiburg: Walter Verlag, 1987

THE BASIC WRITINGS OF C.G. JUNG
New York: The Modern Library, 1959

THE DEVELOPMENT OF PERSONALITY
Collected Writings, Vol. 17
Princeton: Princeton University Press, 1954

THE MEANING AND SIGNIFICANCE OF DREAMS
Boston: Sigo Press, 1991

THE MYTH OF THE DIVINE CHILD
in: Essays on A Science of Mythology
Princeton, N.J.: Princeton University Press Bollingen Series XXII, 1969. (With Karl Kerenyi)

TWO ESSAYS ON ANALYTICAL PSYCHOLOGY
Collected Writings, Vol. 7
Princeton: Princeton University Press, 1972
First published by Routledge & Kegan Paul, Ltd., 1953

## Kahn, Charles (Ed.)

The Art and Thought of Heraclitus
Cambridge: Cambridge University Press, 2008

## Kapleau, Roshi Philip

Three Pillars of Zen
Boston: Beacon Press, 1967

## Karagulla, Shafica

The Chakras
Correlations between Medical Science and Clairvoyant Observation (With Dora van Gelder Kunz)
Wheaton: Quest Books, 1989

## Klein, Melanie

Love, Guilt and Reparation, and Other Works 1921-1945
New York: Free Press, 1984
(Reissue Edition)

Envy and Gratitude and Other Works 1946-1963
New York: Free Press, 2002
(Reissue Edition)

## Kraemer

The Forbidden Love
London, 1976

BIBLIOGRAPHY

## Krafft-Ebing, Richard von

Psychopathia Sexualis
New York: Bell Publishing, 1965
Originally published in 1886

## Krause, Donald G.

The Art of War for Executives
London: Nicholas Brealey Publishing, 1995

## Krishnamurti, J.

Freedom From The Known
San Francisco: Harper & Row, 1969

The First and Last Freedom
San Francisco: Harper & Row, 1975

Education and the Significance of Life
London: Victor Gollancz, 1978

Commentaries on Living
First Series
London: Victor Gollancz, 1985

Commentaries on Living
Second Series
London: Victor Gollancz, 1986

Krishnamurti's Journal
London: Victor Gollancz, 1987

Krishnamurti's Notebook
London: Victor Gollancz, 1986

BEYOND VIOLENCE
London: Victor Gollancz, 1985

BEGINNINGS OF LEARNING
New York: Penguin, 1986

THE PENGUIN KRISHNAMURTI READER
New York: Penguin, 1987

ON GOD
San Francisco: Harper & Row, 1992

ON FEAR
San Francisco: Harper & Row, 1995

THE ESSENTIAL KRISHNAMURTI
San Francisco: Harper & Row, 1996

THE ENDING OF TIME
With Dr. David Bohm
San Francisco: Harper & Row, 1985

## LAING, RONALD DAVID

DIVIDED SELF
New York: Viking Press, 1991

R.D. LAING AND THE PATHS OF ANTI-PSYCHIATRY
ed., by Z. Kotowicz
London: Routledge, 1997

THE POLITICS OF EXPERIENCE
New York: Pantheon, 1983

# BIBLIOGRAPHY

## LAKHOVSKY, GEORGES

SECRET OF LIFE
New York: Kessinger Publishing, 2003

## LASZLO, ERVIN

SCIENCE AND THE AKASHIC FIELD
An Integral Theory of Everything
Rochester: Inner Traditions, 2004

QUANTUM SHIFT TO THE GLOBAL BRAIN
How the New Scientific Reality Can Change Us and Our World
Rochester: Inner Traditions, 2008

SCIENCE AND THE REENCHANTMENT OF THE COSMOS
The Rise of the Integral Vision of Reality
Rochester: Inner Traditions, 2006

THE AKASHIC EXPERIENCE
Science and the Cosmic Memory Field
Rochester: Inner Traditions, 2009

THE CHAOS POINT
The World at the Crossroads
Newburyport, MA: Hampton Roads Publishing, 2006

## LAUD, ANNE & GILSTROP, MAY

VIOLENCE IN THE FAMILY
A Selected Bibliography on Child Abuse, Sexual Abuse of Children & Domestic Violence, June 1985, University of Georgia Libraries, Bibliographical Series, No. 32

## LEADBEATER, CHARLES WEBSTER

### ASTRAL PLANE
Its Scenery, Inhabitants and Phenomena
Kessinger Publishing Reprint Edition, 1997

### DREAMS
What they Are and How they are Caused
London: Theosophical Publishing Society, 1903
Kessinger Publishing Reprint Edition, 1998

### THE INNER LIFE
Chicago: The Rajput Press, 1911
Kessinger Publishing

## LEARY, TIMOTHY

### OUR BRAIN IS GOD
Berkeley, CA: Ronin Publishing, 2001
Author Copyright 1988

## LEBOYER, FREDERICK

### BIRTH WITHOUT VIOLENCE
New York, 1975

### INNER BEAUTY, INNER LIGHT
New York: Newmarket Press, 1997

### LOVING HANDS
The Traditional Art of Baby Massage
New York: Newmarket Press, 1977

# BIBLIOGRAPHY

THE ART OF BREATHING
New York: Newmarket Press, 1991

## LEGGETT, TREVOR P.

A FIRST ZEN READER
Rutland: C.E. Tuttle, 1980
Originally published in 1972

## LEONARD, GEORGE, MURPHY, MICHAEL

THE LIVE WE ARE GIVEN
A Long Term Program for Realizing the
Potential of Body, Mind, Heart and Soul
New York: Jeremy P. Tarcher/Putnam, 1984

## LICHT, HANS

SEXUAL LIFE IN ANCIENT GREECE
New York: AMS Press, 1995

## LIEDLOFF, JEAN

CONTINUUM CONCEPT
In Search of Happiness Lost
New York: Perseus Books, 1986
First published in 1977

## Lipton, Bruce

### The Biology of Belief
Unleashing the Power of Consciousness, Matter and Miracles
Santa Rosa, CA: Mountain of Love/Elite Books, 2005

## Locke, John

### Some Thoughts Concerning Education
London, 1690
Reprinted in: The Works of John Locke, 1823
Vol. IX., pp. 6-205

## Long, Max Freedom

### The Secret Science at Work
The Huna Method as a Way of Life
Marina del Rey: De Vorss Publications, 1995
Originally published in 1953

### Growing Into Light
A Personal Guide to Practicing the Huna Method,
Marina del Rey: De Vorss Publications, 1955

## Lowen, Alexander

### Bioenergetics
New York: Coward, McGoegham 1975

### Depression and the Body
The Biological Basis of Faith and Reality
New York: Penguin, 1992

# BIBLIOGRAPHY

FEAR OF LIFE
New York: Bioenergetic Press, 2003

HONORING THE BODY
The Autobiography of Alexander Lowen
New York: Bioenergetic Press, 2004

JOY
The Surrender to the Body and to Life
New York: Penguin, 1995

LOVE AND ORGASM
New York: Macmillan, 1965

LOVE, SEX AND YOUR HEART
New York: Bioenergetics Press, 2004

NARCISSISM: DENIAL OF THE TRUE SELF
New York: Macmillan, Collier Books, 1983

PLEASURE: A CREATIVE APPROACH TO LIFE
New York: Bioenergetics Press, 2004
First published in 1970

THE LANGUAGE OF THE BODY
Physical Dynamics of Character Structure
New York: Bioenergetics Press, 2006

## MALINOWSKI, BRONISLAW

CRIME UND CUSTOM IN SAVAGE SOCIETY
London: Kegan, 1926

SEX AND REPRESSION IN SAVAGE SOCIETY
London: Kegan, 1927

THE SEXUAL LIFE OF SAVAGES IN NORTH WEST MELANESIA
New York: Halycon House, 1929

## MANN, EDWARD W.

ORGONE, REICH & EROS
Wilhelm Reich's Theory of Life Energy
New York: Simon & Schuster (Touchstone), 1973

## MARTINSON, FLOYD M.

SEXUAL KNOWLEDGE
Values and Behavior Patterns
St. Peter: Minn.: Gustavus Adolphus College, 1966

INFANT AND CHILD SEXUALITY
St. Peter: Minn.: Gustavus Adolphus College, 1973

THE QUALITY OF ADOLESCENT EXPERIENCES
St. Peter: Minn.: Gustavus Adolphus College, 1974

THE CHILD AND THE FAMILY
Calgary, Alberta: The University of Calgary, 1980

THE SEX EDUCATION OF YOUNG CHILDREN
in: Lorna Brown (Ed.), Sex Education in the Eighties
New York, London: Plenum Press, 1981, pp. 51 ff.

THE SEXUAL LIFE OF CHILDREN
New York: Bergin & Garvey, 1994

CHILDREN AND SEX, PART II: CHILDHOOD SEXUALITY
in: Bullough & Bullough, Human Sexuality (1994)
Pp. 111-116

# BIBLIOGRAPHY

## MASTERS, R.E.L.

FORBIDDEN SEXUAL BEHAVIOR AND MORALITY
New York, 1962

## MCCAREY, WILLIAM A.

IN SEARCH OF HEALING
Whole-Body Healing Through the Mind-Body-Spirit Connection
New York: Berkley Publishing, 1996

## MCLEOD, KEMBREW

FREEDOM OF EXPRESSION
Resistance and Repression in the Age of Intellectual Property
Minneapolis, MN: University of Minnesota Press, 2007

## MCTAGGART, LYNNE

THE FIELD
The Quest for the Secret Force of the Universe
New York: Harper & Collins, 2002

## MEAD, MARGARET

SEX AND TEMPERAMENT IN THREE PRIMITIVE SOCIETIES
New York, 1935

## MEADOWS, DONELLA H.

THINKING IN SYSTEMS
A Primer
White River, VT: Chelsea Green Publishing, 2008

## MEHTA, ROHIT

J. KRISHNAMURTI AND THE NAMELESS EXPERIENCE
A Comprehensive Discussion of J. Krishnamurti's Approach to Life
Delhi: Motilal Banarsidass Publishers, 2002

## MERLEAU-PONTY, MAURICE

PHENOMENOLOGY OF PERCEPTION
London: Routledge, 1995
Originally published 1945

## METZNER, RALPH (ED.)

AYAHUASCA, HUMAN CONSCIOUSNESS AND THE SPIRITS OF NATURE
ed. by Ralph Metzner, Ph.D
New York: Thunder's Mouth Press, 1999

THE PSYCHEDELIC EXPERIENCE
A Manual Based on the Tibetan Book of the Dead
With Timothy Leary and Richard Alpert
New York: Citadel, 1995

# BIBLIOGRAPHY

## MILLER, ALICE

FOUR YOUR OWN GOOD
Hidden Cruelty in Child-Rearing and the Roots of Violence
New York: Farrar, Straus & Giroux, 1983

PICTURES OF A CHILDHOOD
New York: Farrar, Straus & Giroux, 1986

THE DRAMA OF THE GIFTED CHILD
In Search for the True Self
translated by Ruth Ward
New York: Basic Books, 1996

THOU SHALT NOT BE AWARE
Society's Betrayal of the Child
New York: Noonday, 1998

THE POLITICAL CONSEQUENCES OF CHILD ABUSE
in: The Journal of Psychohistory 26, 2 (Fall 1998)

## MOLL, ALBERT

THE SEXUAL LIFE OF THE CHILD
New York: Macmillan, 1912
First published in German as
Das Sexualleben des Kindes, 1909

## MONROE, ROBERT

ULTIMATE JOURNEY
New York: Broadway Books, 1994

## Montagu, Ashley

Touching
The Human Significance of the Skin
New York: Harper & Row, 1978

## Montessori, Maria

The Absorbent Mind
Reprint Edition
New York: Buccaneer Books, 1995
First published in 1973

## Moore, Thomas

Care of the Soul
A Guide for Cultivating Depth and Sacredness in Everyday Life
New York: Harper & Collins, 1994

## Moser, Charles Allen

DSM-IV-TR and the Paraphilias: An Argument for Removal
With Peggy J. Kleinplatz
Journal of Psychology and Human Sexuality 17 (3/4), 91-109 (2005)

## Murdock, G.

Social Structure
New York: Macmillan, 1960

# BIBLIOGRAPHY

## Murphy, Joseph

The Power of Your Subconscious Mind
West Nyack, N.Y.: Parker, 1981, N.Y.: Bantam, 1982
Originally published in 1962

The Miracle of Mind Dynamics
New York: Prentice Hall, 1964

Miracle Power for Infinite Riches
West Nyack, N.Y.: Parker, 1972

The Amazing Laws of Cosmic Mind Power
West Nyack, N.Y.: Parker, 1973

Secrets of the I Ching
West Nyack, N.Y.: Parker, 1970

Think Yourself Rich
Use the Power of Your Subconscious Mind to Find True Wealth
Revised by Ian D. McMahan, Ph.D.
Paramus, NJ: Reward Books, 2001

## Murphy, Michael

The Future of the Body
Explorations into the Further Evolution of Human Nature
New York: Jeremy P. Tarcher/Putnam, 1992

## Myers, Tony Pearce

The Soul of Creativity
Insights into the Creative Process
Novato, CA: New World Library, 1999

## Myss, Caroline

The Creation of Health
The Emotional, Psychological, and Spiritual Responses that Promote Health and Healing
New York: Three Rivers Press, 1998

## Naparstek, Belleruth

Your Sixth Sense
Unlocking the Power of Your Intuition
London: HarperCollins, 1998

Staying Well With Guided Imagery
New York: Warner Books, 1995

## Narby, Jeremy

The Cosmic Serpent
DNA and the Origins of Knowledge
New York: J. P. Tarcher, 1999

## Nau, Erika

Self-Awareness Through Huna
Virginia Beach: Donning, 1981

## Neill, Alexander Sutherland

Neill! Neill! Orange-Peel!
New York: Hart Publishing Co., 1972

# BIBLIOGRAPHY

SUMMERHILL
A Radical Approach to Child Rearing
New York: Hart Publishing, Reprint 1984
Originally published 1960

SUMMERHILL SCHOOL
A New View of Childhood
New York: St. Martin's Press
Reprint 1995

## NEUMANN, ERICH

THE GREAT MOTHER
Princeton: Princeton University Press, 1955
(Bollingen Series)

## NEWTON, MICHAEL

LIFE BETWEEN LIVES
Hypnotherapy for Spiritual Regression
Woodbury, Minn.: Llewellyn Publications, 2006

## NICHOLS, SALLIE

JUNG AND TAROT: AN ARCHETYPAL JOURNEY
New York: Red Wheel/Weiser, 1986

## NIN, ANAÏS

THE DIARY OF ANAÏS NIN (7 VOLUMES)
New York, 1966

EIGHT DYNAMIC PATTERNS OF LIVING

VOLUME 1 (1931-1934)
New York: Harvest Books, 1969

VOLUME 2 (1934-1939)
New York: Harvest Books, 1970

## ODENT, MICHEL

BIRTH REBORN
What Childbirth Should Be
London: Souvenir Press, 1994

THE SCIENTIFICATION OF LOVE
London: Free Association Books, 1999

PRIMAL HEALTH
Understanding the Critical Period Between Conception and the First Birthday
London: Clairview Books, 2002
First Published in 1986 with Century Hutchinson in London

THE FUNCTIONS OF THE ORGASMS
The Highway to Transcendence
London: Pinter & Martin, 2009

## OLLENDORF-REICH, ILSE

WILHELM REICH, A PERSONAL BIOGRAPHY
New York, St. Martins Press, 1969

WILHELM REICH
Vorwort von A.S. Neill
München, Kindler, 1975

# BIBLIOGRAPHY

## Pearce Myers, Tony (Editor)

THE SOUL OF CREATIVITY
Insights into the Creative Process
Novato: New World Library, 1999

## Pert, Candace B.

MOLECULES OF EMOTION
The Science Behind Mind-Body Medicine
New York: Scribner, 2003

## Petrash, Jack

UNDERSTANDING WALDORF EDUCATION
Teaching from the Inside Out
London: Floris Books, 2003

## Plummer, Kenneth

PEDOPHILIA
Constructing a Sociological Baseline
in: in: Cook, M. and Howells, K. (Eds.):
Adult Sexual Interest in Children
Academic Press, London, 1980, pp. 220 ff.

## Porteous, Hedy S.

SEX AND IDENTITY
Your Child's Sexuality
Indianapolis: Bobbs-Merrill, 1972

## Prescott, James W.

AFFECTIONAL BONDING FOR THE PREVENTION OF VIOLENT BEHAVIORS
Neurobiological, Psychological and Religious/Spiritual Determinants, in: Hertzberg, L.J., Ostrum, G.F. and Field, J.R., (Eds.)

VIOLENT BEHAVIOR
Vol. 1, Assessment & Intervention, Chapter Six
New York: PMA Publishing, 1990

ALIENATION OF AFFECTION
Psychology Today, December 1979

BODY PLEASURE AND THE ORIGINS OF VIOLENCE
Bulletin of the Atomic Scientists, 10-20 (1975)

DEPRIVATION OF PHYSICAL AFFECTION AS A PRIMARY PROCESS IN THE DEVELOPMENT OF PHYSICAL VIOLENCE A COMPARATIVE AND CROSS-CULTURAL PERSPECTIVE, IN: DAVID G. GIL, ED., CHILD ABUSE AND VIOLENCE
New York: Ams Press, 1979

EARLY SOMATOSENSORY DEPRIVATION AS AN ONTOGENETIC PROCESS IN THE ABNORMAL DEVELOPMENT OF THE BRAIN AND BEHAVIOR,
in: Medical Primatology, ed. by I.E. Goldsmith and J. Moor-Jankowski,
New York: S. Karger, 1971

GENITAL MUTILATION OF CHILDREN: FAILURE OF HUMANITY AND HUMANISM
Unprinted Essay (2005)
http://www.violence.de/prescott/letters/CIRC_CONGRESS_MONTAGUE_9.30.05.html

GENITAL PAIN VS. GENITAL PLEASURE
Why the One and not the Other

BIBLIOGRAPHY

The Truth Seeker, July/August 1989, pp. 14-21
http://www.violence.de/prescott/truthseeker/genpl.html

HOW CULTURE SHAPES THE DEVELOPING BRAIN AND THE FUTURE OF HUMANITY
A Brief Summary of the research which links brain abnormalities and violence to an absence of nurturing and bonding very early in childhood, in: Touch the Future: Optimum Learning Relationships

FOR CHILDREN & ADULTS
Spring 2002 (Ed. by Michael Mendizza)
Nevada City, CA, 2002

INVITED COMMENTARY: CENTRAL NERVOUS SYSTEM FUNCTIONING IN ALTERED SENSORY ENVIRONMENTS
in: M.H. Appley and R. Trumbull (Eds.), Psychological Stress, New York: Appleton-Century Crofts, 1967

OUR TWO CULTURAL BRAINS: NEUROINTEGRATIVE AND NEURODISSOCIATIVE
http://www.violence.de/prescott/letters/Our_Two_Cultural_Brains.pdf

PHYLOGENETIC AND ONTOGENETIC ASPECTS OF HUMAN AFFECTIONAL DEVELOPMENT,
in: Progress in Sexology, Proceedings of the 1976 International, Congress of Sexology, ed. by R. Gemme & C.C. Wheeler, New York: Plenum Press, 1977

PREVENTION OR THERAPY AND THE POLITICS OF TRUST INSPIRING A NEW HUMAN AGENDA
in: Psychotherapy and Politics International
Volume 3(3), pp. 194-211
London: John Wiley, 2005

SEX AND THE BRAIN
Midcontinent & Eastern Regions, June 13-16, 2002

Big Rapids, MI: Society for Cross-Cultural Research, 32nd Annual Meeting, 2005
http://www.violence.de/archive.shtml

SIXTEEN PRINCIPLES FOR PERSONAL, FAMILY AND GLOBAL PEACE
The Truth Seeker, March/April 1989
http://www.violence.de/prescott/letters/Sixteen_Principles.pdf

SOMATOSENSORY AFFECTIONAL DEPRIVATION (SAD) THEORY OF DRUG AND ALCOHOL USE
in: Theories on Drug Abuse: Selected Contemporary Perspectives, ed. by Dan J. Lettieri, Mollie Sayers and Helen Wallenstien Pearson, NIDA Research Monograph 30, March 1980, Rockville, MD: National Institute on Drug Abuse, Department of Health and Human Services, 1980

THE ORIGINS OF HUMAN LOVE AND VIOLENCE
Pre- and Perinatal Psychology Journal, Volume 10, Number 3: Spring 1996, pp. 143-188 The Origins of Love and Violence

SENSORY DEPRIVATION AND THE DEVELOPING BRAIN
Research and Prevention (DVD)
http://ttfuture.org/store/origins_orders
http://violence.de
http://ttfuture.org/violence
http://montagunocircpetition.org

## PRITCHARD, COLIN

THE CHILD ABUSERS
New York: Open University Press, 2004

## RAKNES, OLA

WILHELM REICH AND ORGONOMY
Oslo: Universitetsforlaget, 1970

## RANDALL, NEVILLE

LIFE AFTER DEATH
London: Robert Hale, 1999

## RANK, OTTO

ART AND ARTIST
With Charles Francis Atkinson and Anaïs Nin
New York: W.W. Norton, 1989
Originally published in 1932

THE SIGNIFICANCE OF PSYCHOANALYSIS FOR THE MENTAL SCIENCES
New York: BiblioBazaar, 2009
First published in 1913

## REDFIELD, JAMES

THE TENTH INSIGHT
Holding the Vision
New York: Warner Books, 1996

THE CELESTINE PROPHECY
New York: Warner Books, 1995

# EIGHT DYNAMIC PATTERNS OF LIVING

## REICH, WILHELM

A REVIEW OF THE THEORIES, DATING FROM THE 17TH CENTURY, ON THE ORIGIN OF ORGANIC LIFE
by Arthur Hahn, Literature Assistant at the Institut für Sexualökonomische Lebensforschung, Biologisches Laboratorium, Oslo, 1938, ©1979 Mary Boyd Higgins as Director of the Wilhelm Reich Infant Trust, XEROX Copy from the Wilhelm Reich Museum

CHILDREN OF THE FUTURE
On the Prevention of Sexual Pathology
New York: Farrar, Straus & Giroux, 1984
First published in 1950

CORE (COSMIC ORGONE ENGINEERING)
Part I, Space Ships, DOR and DROUGHT
©1984, Orgone Institute Press
XEROX Copy from the Wilhelm Reich Museum
Köln: Kiepenheuer & Witsch, 1987

EARLY WRITINGS 1
New York: Farrar, Straus & Giroux, 1975

ETHER, GOD & DEVIL & COSMIC SUPERIMPOSITION
New York: Farrar, Straus & Giroux, 1972
Originally published in 1949

GENITALITY IN THE THEORY AND THERAPY OF NEUROSIS
©1980 by Mary Boyd Higgins as Director of the Wilhelm Reich Infant Trust

PEOPLE IN TROUBLE
©1974 by Mary Boyd Higgins as Director of the Wilhelm Reich Infant Trust

# BIBLIOGRAPHY

RECORD OF A FRIENDSHIP
The Correspondence of Wilhelm Reich and A. S. Neill
New York, Farrar, Straus & Giroux, 1981

SELECTED WRITINGS
An Introduction to Orgonomy
New York: Farrar, Straus & Giroux, 1973

THE BIOELECTRICAL INVESTIGATION OF SEXUALITY AND ANXIETY
New York: Farrar, Straus & Giroux, 1983
Originally published in 1935

THE BION EXPERIMENTS
reprinted in Selected Writings
New York: Farrar, Straus & Giroux, 1973

THE CANCER BIOPATHY (THE ORGONE, VOL. 2)
New York: Farrar, Straus & Giroux, 1973

THE FUNCTION OF THE ORGASM (THE ORGONE, VOL. 1)
Orgone Institute Press, New York, 1942

THE INVASION OF COMPULSORY SEX MORALITY
New York: Farrar, Straus & Giroux, 1971
Originally published in 1932

THE LEUKEMIA PROBLEM: APPROACH
©1951, Orgone Institute Press
Copyright Renewed 1979
XEROX Copy from the Wilhelm Reich Museum

THE MASS PSYCHOLOGY OF FASCISM
New York: Farrar, Straus & Giroux, 1970
Originally published in 1933

THE ORGONE ENERGY ACCUMULATOR
Its Scientific and Medical Use

©1951, 1979, Orgone Institute Press
XEROX Copy from the Wilhelm Reich Museum

THE SCHIZOPHRENIC SPLIT
©1945, 1949, 1972 by Mary Boyd Higgins as Director of the Wilhelm Reich Infant Trust
XEROX Copy from the Wilhelm Reich Museum

THE SEXUAL REVOLUTION
©1945, 1962 by Mary Boyd Higgins as Director of the Wilhelm Reich Infant Trust

## RISO, DON RICHARD & HUDSON, RUSS

THE WISDOM OF THE ENNEAGRAM
The Complete Guide to Psychological and Spiritual Growth For The Nine Personality Types
New York: Bantam Books, 1999

## ROBBINS, ANTHONY

AWAKEN THE GIANT WITHIN
New York: Simon & Schuster, 1991

UNLIMITED POWER
The New Science of Personal Achievement
New York: Free Press, 1997

## ROBERTS, JANE

THE NATURE OF PERSONAL REALITY
New York: Amber-Allen Publishing, 1994
First published in 1974

## BIBLIOGRAPHY

THE NATURE OF THE PSYCHE
Its Human Expression
New York, Amber-Allen Publishing, 1996
First published in 1979

## ROSEN, SYDNEY (ED.)

MY VOICE WILL GO WITH YOU
The Teaching Tales of Milton H. Erickson
New York: Norton & Co., 1991

## ROTHSCHILD & WOLF

CHILDREN OF THE COUNTERCULTURE
New York: Garden City, 1976

## SANDFORT, THEO

THE SEXUAL ASPECT OF PEDOPHILE RELATIONS
The Experience of Twenty-five Boys
Amsterdam: Pan/Spartacus, 1982

## SCHLIPP, PAUL A. (ED.)

ALBERT EINSTEIN
Philosopher-Scientist
New York: Open Court Publishing, 1988

## SCHWARTZ, ANDREW E.

**GUIDED IMAGERY FOR GROUPS**
Fifty Visualizations That Promote Relaxation, Problem-Solving, Creativity, and Well-Being
Whole Person Associates, 1995

## SHARAF, MYRON

**FURY ON EARTH**
A Biography of Wilhelm Reich
London: André Deutsch, 1983

## SHELDRAKE, RUPERT

**A NEW SCIENCE OF LIFE**
The Hypothesis of Morphic Resonance
Rochester: Park Street Press, 1995

## SHER, BARBARA & GOTTLIEB, ANNIE

**WISHCRAFT**
How to Get What You Really Want
2nd edition, New York: Ballantine Books, 2003

## SHONE, RONALD

**CREATIVE VISUALIZATION**
Using Imagery and Imagination for Self-Transformation
New York: Destiny Books, 1998

# BIBLIOGRAPHY

## SIMONTON, O. CARL ET AL.

GETTING WELL AGAIN
Los Angeles: Tarcher, 1978

## SINGER, JUNE

ANDROGYNY
New York: Doubleday Dell, 1976

## SMITH, C. MICHAEL

JUNG AND SHAMANISM IN DIALOGUE
London: Trafford Publishing, 2007

## SPOCK, BENJAMIN

DR. SPOCK'S BABY AND CHILD CARE
8th Edition
New York: Pocket Books, 2004

## STEIN, ROBERT M.

REDEEMING THE INNER CHILD IN MARRIAGE AND THERAPY
in: Reclaiming the Inner Child
ed. by Jeremiah Abrams
New York: Tarcher/Putnam, 1990, 261 ff.

## Steiner, Rudolf

Theosophy
An Introduction to the Spiritual Processes in Human Life and in the Cosmos
New York: Anthroposophic Press, 1994

## Stekel, Wilhelm

Auto-Eroticism
A Psychiatric Study of Onanism and Neurosis
Republished, London: Paul Kegan, 2004

Patterns of Psychosexual Infantilism
New York, 1959 (reprint edition)

Sadism and Masochism
New York: W.W. Norton & Co., 1953

Sex and Dreams
The Language of Dreams
Republished
New York: University Press of the Pacific, 2003

## Stiene, Bronwen & Frans

The Reiki Sourcebook
New York: O Books, 2003

The Japanese Art of Reiki
A Practical Guide to Self-Healing
New York: O Books, 2005

# BIBLIOGRAPHY

## STONE, HAL & STONE, SIDRA

EMBRACING OUR SELVES
The Voice Dialogue Manual
San Rafael, CA: New World Library, 1989

## STRASSMAN, RICK

DMT: THE SPIRIT MOLECULE
A doctor's revolutionary research into the biology of near-death and mystical experiences
Rochester: Park Street Press, 2001

## SYMONDS, JOHN ADDINGTON

A PROBLEM IN GREEK ETHICS
New York: M.S.G. House, 1971

## SZASZ, THOMAS

THE MYTH OF MENTAL ILLNESS
New York: Harper & Row, 1984

## TALBOT, MICHAEL

THE HOLOGRAPHIC UNIVERSE
New York: HarperCollins, 1992

## Tarnas, Richard

**Cosmos and Psyche**
Intimations of a New World View
New York: Plume, 2007

**The Passion of the Western Mind**
Understanding the Ideas that have Shaped Our World View
New York: Ballantine Books, 1993

## Tart, Charles T.

**Altered States of Consciousness**
A Book of Readings
Hoboken, N.J.: Wiley & Sons, 1969

## Textor, R. B.

**A Cross-Cultural Summary**
New Haven, Human Relations Area Files (HRAF) Press, 1967

## The Advent of Great Awakening

**A Course in Miracles**
Text Workbook and Manual for Teachers
New York: New Christian Church of Full Endeavor, 2007

# BIBLIOGRAPHY

## Tiller, William A.

Conscious Acts of Creation
The Emergence of a New Physics
Associated Producers, 2004 (DVD)

Psychoenergetic Science
New York: Pavior, 2007

## Toffler, Alvin

Powershift
Knowledge, Wealth, and Violence at the Edge of the 21st Century
New York: Bantam, 1991

Revolutionary Wealth
How it will be created and how it will change our lives
New York: Broadway Business, 2007

The Third Wave
New York: Bantam, 1984

## Tolle, Eckhart

The Power of Now
A Guide to Spiritual Enlightenment
Novato, CA: New World Library, 2004

A New Earth: Awakening to Your Life's Purpose
New York: Michael Joseph (Penguin), 2005

## Van Gelder, Dora

The Real World of Fairies
A First-Person Account
2nd Edition
Wheaton: Quest Books, 1999

## Villoldo, Alberto

Healing States
A Journey Into the World of Spiritual Healing and Shamanism
With Stanley Krippner
New York: Simon & Schuster (Fireside), 1987

Dance of the Four Winds: Secrets of the Inca Medicine Wheel
With Eric Jendresen
Rochester: Destiny Books, 1995

Shaman, Healer, Sage
How to Heal Yourself and Others with the Energy Medicine
of the Americas
New York: Harmony, 2000

Healing the Luminous Body
The Way of the Shaman with Dr. Alberto Villoldo
DVD, Sacred Mysteries Productions, 2004

Mending The Past And Healing The Future with Soul Retrieval
New York: Hay House, 2005

## Whitfield, Charles L.

Healing the Child Within
Deerfield Beach, Fl: Health Communications, 1987

## WHITING, BEATRICE B.

CHILDREN OF SIX CULTURES
A Psycho-Cultural Analysis
Cambridge: Harvard University Press, 1975

## WILBER, KEN

SEX, ECOLOGY, SPIRITUALITY
The Spirit of Evolution
Boston: Shambhala, 2000

QUANTUM QUESTIONS
Mystical Writings of The World's Greatest Physicists
Boston: Shambhala, 2001

## WILLIAMS, STREPHON KAPLAN

DREAMS AND SPIRITUAL GROWTH
With Patricia H. Berne and Louis M. Savary
New York: Paulist Press, 1984

DREAM CARDS
Understand Your Dreams and Enrich Your Life
New York: Simon & Schuster (Fireside), 1991

## WOLF, FRED ALAN

TAKING THE QUANTUM LEAP
The New Physics for Nonscientists
New York: Harper & Row, 1989

PARALLEL UNIVERSES
New York: Simon & Schuster, 1990

THE DREAMING UNIVERSE
A Mind-Expanding Journey into the Realm Where Psyche and Physics Meet
New York: Touchstone, 1995

THE EAGLE'S QUEST
A Physicist Finds the Scientific Truth At the Heart of the Shamanic World
New York: Touchstone, 1997

## YATES, ALAYNE

SEX WITHOUT SHAME: ENCOURAGING THE CHILD'S HEALTHY SEXUAL DEVELOPMENT
New York, 1978
Republished Internet Edition

## ZUKAV, GARY

THE DANCING WU LI MASTERS
An Overview of the New Physics
New York: HarperOne, 2001

# Personal Notes